PRAISE FOR
DEFY

"Dr. Sunita Sah shows us that defiance isn't just for protests; it's just as vital at the kitchen table. Even though I'm someone who is dedicated to challenging the system, *Defy* showed me how often I let others' beliefs eclipse my own—diminishing my sense of agency. Dr. Sah powerfully argues that the ability to defy is essential for true liberation, even from self-imposed constraints. *Defy* is a must-read for anyone committed to justice and building an inclusive world."

—**LaTosha Brown**, organizer, strategist, and co-founder
  of Black Voters Matter

"We navigate the world by choosing what we will or will not accept but Dr. Sunita Sah is delivering, for the first time, the science behind those choices. What do you do when you're faced with a big decision? What do you do when you feel pressured to make a particular choice? From your best yes to your truest no, *Defy* helps us handle these decisions with integrity, compassion, and candor."

—**Annie Duke**, bestselling author of *Quit, How to Decide,*
  and *Thinking in Bets*

"Dr. Sah has created this decisive how-to guide on defiance and acting in accordance with your deepest convictions. Firmly grounded in the latest science, *Defy* will surely become a culture-shifting manifesto for all of us 'moral mavericks' who hope to use conscientious resistance to make the world a better place."

—**Jonah Berger**, *New York Times* bestselling author of *Magic Words,*
  The Catalyst, Contagious, and *Invisible Influence*

"Dr. Sah's *Defy* is the answer we need for all those regret-filled moments when we stay silent and close our eyes. A playbook to understand how and when to take action, *Defy* is an essential and transformative read for everyone wanting to live a life true to their values."

—**Dr. Mona Hanna**, author of *What the Eyes Don't See*

"As parents, we are the front lines of defense for our children's well-being. We inevitably have to contend with the advice of other family members, teachers, and medical professionals, and along with that, maintain trust in our children. Dr. Sah's work on how we can learn to conscientiously defy is a must-read for any parent, woman, or professional."

—**Tina Payne Bryson**, *New York Times* bestselling co-author of *The Whole-Brain Child, No-Drama Discipline*, and *The Way of Play*

"*Defy* answers one of the most pressing questions raised by the pandemic, protests, and the presidential election: why do people behave in ways that ignore science, oppose equity, and undercut their own self-interest? Dr. Sah's ability to explain how we are both hardwired to comply and socially conditioned to obey provides much-needed insight on everything from police brutality to anti-mask/anti-vax protests, to the Capitol insurrection. But *Defy* isn't a dry tome focusing on scholarly analysis of current events. It's an engaging explanation of why defiance is the exception and obedience is the rule in all of our lives while also providing concrete strategies to recognize and transform our responses."

—**Kerry Ann Rockquemore**, founder of the National Center for Faculty Development & Diversity

"Dr. Sah's *Defy* is an important, timely book written by an expert who has studied and lived the central questions in her book: when should you speak up, and how do you empower people at home, work, and play to speak up when staying silent is the easier option? *Defy* is an important book for policymakers, educators, parents, and really anyone who's faced the prospect of standing up to power, indifference, and callousness."

—**Adam Alter**, *New York Times* bestselling author of *Irresistible* and *Drunk Tank Pink*

DEFY

DEFY

**The Power of No in a
World That Demands Yes**

DR. SUNITA SAH

ONE WORLD
NEW YORK

Published in the United States by One World, an imprint of Random House,
a division of Penguin Random House LLC, New York.

ONE WORLD and colophon are registered trademarks of
Penguin Random House LLC.

Hardback ISBN 978-0-593-44577-8
Ebook ISBN 978-0-593-44578-5

Printed in the United States of America on acid-free paper

oneworldlit.com

2 4 6 8 9 7 5 3 1

FIRST EDITION

To my parents, who showed me that defiance is not just an act
but a testament to our deepest values.

Sometimes we are blessed with being able to choose the time, and the arena, and the manner of our revolution, but more usually we must do battle where we are standing.

Audre Lorde

CONTENTS

Old Definition (*Oxford English Dictionary,* 2024)

- -

defy, verb.

to challenge the power of; to resist boldly or openly

New Definition (Sunita Sah, 2025)

- -

defy, verb.

to act in accordance with your true values when there is pressure to do otherwise

INTRODUCTION

Moving the Knee

When you think of the long and gloomy history of man,
you will find more hideous crimes have been committed
in the name of obedience than have ever been
committed in the name of rebellion.

C. P. Snow

On May 25, 2020, two rookie police officers in Minneapolis responded to a call about a man using a counterfeit bill at a convenience store.

When they arrived at the scene, they found the suspect, a Black man in his forties, sitting in the driver's seat of a car. One rookie drew his gun on the man and forced him out of the vehicle. They handcuffed him.

Two more police officers arrived. They pinned him to the ground, and the most experienced cop at the scene, a man named Derek Chauvin, put his knee on the back of the man's neck.

Most people know the rest of this story:

How George Floyd, face down on the ground, said again and again that he could not breathe.

How he cried out in pain and lost consciousness.

How Chauvin kept his knee on Floyd's neck for over nine minutes—even after Floyd showed no signs of life, even after the paramedics arrived.

George Floyd's death lays bare the worst of systemic racism and police brutality. Some do not like to remember what happened, while others want to make sure we do not forget.

What truly haunts me, when I watch the footage of that day, isn't just the sight of the prone man on the ground or the horrific image of the knee on the back of his neck.

It's also the two rookie cops—one white, on his fourth day on the job; one Black, on his third day; one holding Floyd's legs, the other crouched over his back—who follow Chauvin's lead while a crowd of horrified bystanders begs Chauvin to remove his knee from George Floyd's neck.

The white rookie was Thomas Lane. He was thirty-seven years of age; his grandfather had been a homicide detective, and his great-grandfather the Minneapolis chief of police.

The Black rookie was Alex Kueng. The biracial son of a white mother and a Black father, and more than a decade younger than Lane, Kueng had joined the police department with hopes of reforming it from the inside. Many people he knew—including his four adopted siblings, one of whom had previously been arrested for filming police officers and threatened with a Taser—distrusted the police. Kueng watched for most of his life as his community's relationship with the police frayed. He told his family and friends that by joining the department, he could be a bridge. He could help protect people like his siblings, people like him—people like George Floyd.

But on that day in May, both rookie officers complied with Derek Chauvin's orders to restrain George Floyd. The footage chills me. Kueng checks Floyd's pulse, twice, and finding none, continues to obey Chauvin. Lane asks, twice, whether they should move Floyd to his side, but when told no, continues to obey Chauvin.

These two officers were the closest people to Derek Chauvin's knee. They could have moved it. They could have told Chauvin to stop. They could have asked if it was really necessary to keep pinning down a man who had stopped resisting long ago.

But they didn't.

As I watched the footage and reporting in the weeks that followed, I kept wondering:

Why, in that moment, did Alex Kueng lose track of his stated purpose for joining the force? Why did he disconnect from his values?

Why did Thomas Lane, who asked Chauvin twice if they should turn Floyd on his side—a simple act that could have saved his life—not escalate his concerns, and say *Stop*?

Did either of them, at any point, realize something was very wrong—and that by being compliant, they were in fact failing to stop an unfolding murder?

I'm a physician turned organizational psychologist; I have spent decades studying ethics, advice, and influence. Much of my research centers on why and when people accept or reject certain instructions, orders, or advice. And after the killing of George Floyd, I kept asking myself:

What would I have done?

Like most people, I believe that I am a moral person. I do not think it is right to kill a man in the street. I believe that police officers should protect people, not commit acts of violence against them.

I'd like to think that if I were in those officers' shoes, I would have acted differently that day. I'd like to think that I would have spoken up.

That I would have said *Stop*.

But I have studied defiance and authority for years. And one thing I have learned is that despite everything we believe about ourselves, we often go along with things even when we know we shouldn't, in situations both life-threatening and trivial. We are so conditioned to comply—especially in certain contexts—that we might not even realize that the moment calls for defiance until it is too late. We believe we'll do the right thing in the circumstances, but then we freeze, confused and unprepared. We fail to put our values into action.

Like most people, I like to think that I'm the sort of person who would have moved Derek Chauvin's knee off of George Floyd's neck.

But am I?

A few years before that day in Minneapolis, when I was living in Pittsburgh, I visited an emergency room for sudden onset chest pain, a deep ache in the center of my chest. I had not experienced this pain before, and I was worried.

Within minutes, I was through triage, a nurse whisking me past the swinging doors and into an examination room. The doctor was a young and confident white woman; her pristinely pressed white coat and neatly styled short black hair matched her serious expression and no-nonsense manner. She listened to my heart and my breathing, then conducted an electrocardiogram, noting nothing out of the ordinary. I was relieved and told her that my pain was subsiding. I expected her to discharge me, but instead, she told me I had to have a CT scan before she could let me leave.

"Why?" I asked.

"Just to be sure you don't have a pulmonary embolism," she said.

With my medical training, I knew that a pulmonary embolism—a blood clot in the lungs—causes a sharp, stabbing pain in the chest, one that catches your breath as you inhale and exhale. We call it "pleuritic chest pain," and I was not experiencing it. I was certain that the scan was unnecessary: I didn't have what it would be looking for. And a CT scan would expose me to ionizing radiation—about seventy times more on average than an X-ray. This is still a small amount, but any exposure to radiation can increase the risk of cancer in the future—why take this chance for no real reason?

Also, as a scholar of decision-making and medical ethics, I am familiar with how costly and harmful overdiagnosis and overtreatment can be. I know that sometimes "less is more" in medicine. I also agree with the bedrock principles of medical ethics, including *nonmaleficence* (doing no harm) and *autonomy* (freedom to choose). I believe that patients should not submit to unnecessary procedures or treatments just because "it's the way things are done."

I should have said no.

But I was lying on a hospital bed in a chilly room, staring up at the fluorescent lights. The doctor wasn't asking for my opinion. She was telling me what she was going to do. And before I could object, she had disappeared and someone else had wheeled me into the scanning room. Ahead of me was the plastic tunnel of the CT scanner, with its raised platform. I sat down on it.

"It's only a small amount of radiation, right?" I asked the imaging technician as she was setting up the equipment, even though I already knew the answer.

"Well . . ." she hedged, with some hesitation in her voice, before continuing with her instructions.

I didn't want the scan. I didn't need the scan.

But I also didn't want the medical team to think I didn't trust their judgment. I didn't want to make even a minor scene.

So when the technician asked me to, I lay back on the gantry, and the machine started up.

As the scanner's tube whirred and clicked, I was perplexed.

I knew that the scan I was undergoing was unnecessary. I knew it didn't align with my values. I knew it could potentially be harmful.

And yet I had agreed to it. Despite my knowledge and better judgment, I had gone along with what the doctor ordered.

Why?

Because she told me to.

Years later, watching the news coverage of George Floyd's death unfold, I thought of that moment.

I couldn't even defy the order of a smiling woman in a white coat.

What makes me think I could stop a man with a gun?

There is of course an enormous difference between undergoing an unnecessary CT scan and allowing a man to be killed. My failure to speak up resulted in a small dose of radiation that affected only myself. Alex Kueng's and Thomas Lane's compliance contributed to the unnecessary death of another human being.

I cannot know exactly what passed through the rookie officers' minds during those nine minutes and twenty-nine seconds that George Floyd was on the ground. But in countless situations like theirs, and mine, and many in between, we comply before we have time to think at all. If we do have a moment, we may ask ourselves:

Do I go along with this, even though it doesn't feel right?

And in too many situations, with consequences minor and major, in decisions that affect others and those that affect only ourselves, the unfortunate answer is:

Yes.

This may be shocking, but as an organizational psychologist, I do not find it surprising. My research has shown me, again and again, that what someone *believes* their values to be is quite different from how they actually *behave.* In experiment after experiment, I've learned that for many of us, the distance between who we *think* we are and what we actually *do* is enormous.

And my research shows that it is in fact incredibly difficult for most people to defy an order, even an unspoken one—from an authority, a peer, or even a stranger. Despite our ideals and best intentions, we most often choose obedience over disobedience, submission over disruption, compliance over defiance.

This is true *regardless of the stakes,* regardless of whether the situation is life-threatening or prosaic. The constant between my situation and the rookie cops—one small and personal, the other major and with ripples of devastation that ran through an entire family and nation—is that critical *moment of compliance.* And that moment happens everywhere—in every workplace, every home, every classroom; it repeats itself across every arena of society, from our intimate relationships to our systems of government.

Defiance is the exception. Obedience is the rule.

But it doesn't have to be that way.

. . .

I moved to the United States from the United Kingdom in 2008, and the years since then have seen some of the most sustained civic protests in American history. From Occupy Wall Street to Black Lives Matter; from Colin Kaepernick to Emma González; from the Women's March to Standing Rock to the COVID-19 rallies against mask mandates—the years I have spent in the United States have been notable for their scenes of open resistance.

People have marched, they have carried banners, they have gone on strike. And they have done so in the face of an increasingly authoritarian government, a burgeoning surveillance state, and a highly polarized political environment. Pressures both legal and social have made the act of rebellion ever more fraught—and, for some, ever more necessary.

Broad social movements can make a difference. But what about small-scale defiance?

Every protest is made up of thousands of individual marchers; every picket line has hundreds of individual signs. Each of those individuals is making a conscious choice to defy: to stand up for what they believe in, despite the costs.

As a psychologist, I am interested in the motivations and abilities of these individual actors. What allows them to overcome their reluctance or outside pressure? What makes them disregard the social costs, the material consequences, even the risk to their physical safety? What, in other words, is going on in our heads when we decide to defy?

What is going through the mind of my teenage son when I tell him to do his homework before turning on his Xbox . . . and then he does the *exact* opposite?

Is an action really defiant if it relies on the beliefs or wishes of someone else?

How can our impulse to defy be used against us?

And, most importantly:

How do we decide when to say yes, and when to say no?

The answers to these questions aren't only found on the streets, in the heads of revolutionaries and iconoclasts, or within the halls of academia.

They are within all of us. Because for every broad social moment of defiance, there are thousands of smaller instances, interactions of minor defiance whose impact isn't felt—at least not immediately—beyond our own lives.

Maybe it's a teenager refusing to participate in playground bullying.

Maybe it's the only female attorney in the firm declining to take on more work than her male peers.

Maybe it's a young person challenging a family member for making a bigoted remark at the dinner table.

Such moments don't appear in the history books, but that doesn't mean they aren't real, or that those small decisions to defy can't have huge societal impact.

The truth is, every social movement begins with one individual choice. And we all face choices like the one Alex Kueng and Thomas Lane confronted that fateful day in Minneapolis. In our families, our jobs, and our relationships; in our homes and our offices; on the streets and in school board meetings—all of us encounter situations every day that force us to decide:

Do I comply?

Or do I defy?

How we answer these questions doesn't only change our own lives.

It can also change the world.

This book is about learning to defy. It's about how to live a life aligned with your values. It's about how to avoid finding yourself complicit in a situation that drastically opposes your principles. It's about what it means to say yes, what it means to say no, and what happens in that critical moment when we have to make a choice between the two.

As a scientist, I like to begin by defining my terms. So before we go any further, I want to explain what I mean by defiance.

When most people hear the word "defy," they think of disobedience. The *Oxford English Dictionary* (*OED*) defines *defy* as "to resist boldly or openly" and *disobedience* as the "refusal to obey" and "acting in defiance."

I am not usually one to disagree with the *OED*. I was raised in the U.K., after all. But as a social scientist and psychologist, I believe this definition is too narrow in scope.

I'd like to propose a new definition of defiance, one that honors our individual agency and our considered actions.

Defiance means *acting in accordance with your true values when there is pressure to do otherwise.*

It sounds simple, but as years of research confirm, it is so much easier said than done.

Many of us don't think deeply about what our values actually are unless we are explicitly asked to. Values are not facts, beliefs, goals, or objectives. They do not depend on our political affiliations or religion, and they do not change when we are in a different situation with different people. Values often distill into single words, such as *honesty, equality,* and *compassion.* They are words that have great meaning and tremendous power if only we could enact them every day.

But our societies are not set up for defiance. We are not always encouraged—or "allowed"—to act in accordance with our values. We are taught from a young age how to obey, and that obedience is good, disobedience bad. We are not given similar instructions for how to defy.

So, all too often, we give up our agency without thinking. We say yes when we don't mean it. We don't say no when we should.

We *actively resist defiance.*

In my research, I have found three key reasons for why this happens:

1. We face enormous pressure to go along with what others want us to do, whether they are authority figures, friends, family, or even strangers.

2. Most of us don't really understand what compliance and defiance are.

3. Once we decide to defy, to act on our true values, we don't know
    how to go about it. We lack the ability to translate our inner
    defiance into outer action.

As a result, our lives are filled with quiet compliance, moments of almost unthinking acceptance of the status quo. I say "almost" because
most of us feel some discomfort about these moments, when our obedience or acceptance is assumed. I felt it in the doctor's exam room when I
accepted a scan I didn't want. Perhaps the rookie cops felt it as well, as
those nine minutes ticked by: a feeling that something was wrong, but
they lacked the tools to change it.

You've probably felt it, too: a kind of powerlessness, a reluctant inability to act according to your own values, a yes that doesn't quite mean *yes*.
It may show up as tension in the back of your neck, a headache, a stomach
cramp, sweat, or a general uneasiness. It's natural to want to get rid of that
feeling or ignore it.

And yet, that discomfort is the key to our power to defy. It is an internal compass, guiding us back to our values. But we often disregard it in
favor of broad obedience—to authority figures, cultural mores, or prevail-
ing social norms.

The forces that lead to compliance are more complex than they might
appear, but they are not insurmountable. We may not always know how
to defy.

But we can learn.

We often think of defiance as LOUD, **bold**, and underlined—or as violent, angry, and aggressive—or as heroic, superhuman, and out of reach.
But the truth is, it need not be any of these things. Defiance also has a
quieter, small-scale side—which can have enormous impacts on our lives
and the lives of the people around us. It isn't only for the brave, or the
extraordinary: it's available, and necessary, for all of us. We can defy in
ways that are the most comfortable and natural for us, with less anxiety

and angst. And the more we practice, the more we prepare ourselves for the critical moment when it arrives.

I will share what I've learned from numerous scientific studies as well as my own personal journey: training to transform myself from a "good girl" who goes along to get along to someone who can defy when needed. I'll also share stories of people from all walks of life who have defied in unexpected ways, sometimes against great odds.

The costs of defiance can be significant. We may lose our jobs, our friends, and the support of our community—and defiance has different stakes for different groups of people. Women, people of color, and other marginalized groups with less power often find themselves in situations in which they need to defy more frequently. Yet, they also face disproportionate consequences for their defiance.

But, as so many of the stories here show, the costs of compliance can be high as well. This shows us that defiance is something we *all* need to do. When we discard our values, it affects us both psychologically and physically, corroding our sense of self. But if we can break free from the culture of compliance that surrounds us, we can start to create new possibilities for ourselves and our communities. Ultimately, the freedom to defy is the freedom to be our true selves.

When I began writing this book, someone asked me:

What do you want people to do after reading your book?

It's a good question, and all of the answers they initially suggested didn't quite fit:

Start a revolution?

Maybe, if the situation warrants it.

Take to the streets?

When necessary.

Create safer and better work environments and more satisfying personal relationships?

Definitely.

But what I'm most interested in is the first domino to fall, the tipping point of every protest no matter how small: the individual *decision* to defy, with all of its complicating factors—psychological, cultural, social, and political.

Defiance doesn't mean living your life in a constant state of acrimonious rebellion. Nor does it require grand actions. It does not even require confrontation.

You don't have to be Mahatma Gandhi or Cesar Chavez, Marsha P. Johnson or Malala Yousafzai. You don't have to be active in a political cause. And you certainly don't always have to defy.

You just have to dare.

Dare to imagine a life lived in accordance with your actual values. Dare to reclaim your agency. Dare to make decisions that align with your innermost principles, no matter the external pressures.

This book isn't about organizing. It isn't a set of instructions on civil disobedience. It isn't a history of protest movements, or an analysis of current social and political activism.

It's about what to do when you're in an exam room being told to get unnecessary tests.

It's about how to respond when you find out that you're paid far less than your co-worker for doing the same job—and how to assess the risks and the rewards of doing so.

It's about how to stand up for yourself and others, when the rest of the room is silent.

It's about how to turn your best intentions into action, whatever the opposition.

It's about how to live your one and only life aligned with your values.

It's about learning how to move the knee.

Part One

A TRUE YES

1

Wired to Comply

n 2012, when the Summer Olympic Games were held in London, I was eager to see the flame pass by on its way to the stadium and the opening ceremonies. On the day of the processional, I arranged to meet my husband on a street corner a short walk away from our apartment, with our then five-year-old son in tow. I had in mind a pleasant afternoon stroll, followed by an exciting glimpse of the torch, held aloft by one of the athletes.

"This will be fun," I told my son, who didn't look convinced. "You're going to witness history!"

Although the walk started well, within a few minutes it became apparent that getting all the way there would be a struggle. The day was hot and he was tired. Before our apartment building was even out of sight, his heels were dragging. A few minutes after that, he asked to be carried. When I refused—he was too heavy—he sat down firmly on the pavement.

"I don't want to go," he told me, his large brown eyes squinting in the sun.

I explained that the Olympics were special. I told him that seeing the famous flame in our neighborhood was a "once-in-a-lifetime experience." But he was unmoved.

"I don't want to see it," he answered, jutting his chin out. "I want to go home."

I pulled on his arm. I asked him firmly. I even tried to pick him up. But I couldn't get more than a few steps down the stifling, crowded pavement before setting him back down again. I was frazzled, hot, and frustrated. We were going to miss the flame.

"Why can't you be good?" I said to my son as we walked home.

My son's only answer was a mulish shrug.

We never saw the flame. I watched it on television later that evening, and my husband laughed when I related the episode to him.

"You sound just like one of those people," he said. "Remember them?"

I did. Soon after my son was born, I had been frequently puzzled by a question well-meaning acquaintances would ask:

Is he good?

What they meant was: Does your baby sleep when you want him to? Does he stop crying when you want him to?

In other words, does he do what *you* want him to do? Does he do what he's told?

As someone fascinated by defiance, this moral equation of obedience to goodness always perplexed me. I had spent years questioning and resisting it—not only as an academic, but as a child of a strict upbringing. I sometimes joked that I studied defiance because my childhood had already given me a world-class education in how to be obedient.

And yet that London afternoon found me standing in the hot sun, pleading with my stubborn son to be "good"—to do what *I* wanted him to do.

That's what kept me up that night, long after the opening ceremonies were over and the rest of my family was asleep. I should have known better—should have known that I had just repeated to him the most basic, unexamined equation that people had impressed on me growing up:

Compliance = good.
Defiance = bad.

From a young age, we are taught to obey.

The first authority figure we routinely encounter is a caregiver, usually a parent, someone whose job it is to nourish us and help us survive, and whose instructions must be followed. Later, teachers often step into the picture, instructing their students not only how to read and do simple arithmetic, but also how to follow the social protocols of the classroom: sit still, raise your hand. Then—as anyone who has attended middle school knows—there is the considerable pressure from our peers to do things the way everyone else is doing them.

This early training has a large effect on us: psychologically, socially, and even neurologically. When we are young, our brains grow at unprecedented rates, forming neural connections and structures that will affect how we behave for years to come. Obedience isn't just a survival technique; it affects our wiring and quite literally shapes our brains. When we are rewarded for compliant behavior, our brain's level of dopamine—the neural transmitter that facilitates our experience of pleasure, among other things—rises. If we consistently repeat those behaviors, we build and strengthen neural pathways for compliance. Unrewarded, disobedient behavior doesn't give our brains the same dopamine rush, and as a result, those behaviors are less likely to be repeated, and those pathways weaken or fail to develop.

We also learn many behaviors through imitation of our parents and caregivers, of our teachers and other authority figures, and of our peers. Psychologically, the value of belonging to a group—mirroring behaviors, attitudes, and actions—becomes apparent before most of us are out of kindergarten, and it continues to shape our behavior as we grow older and enter the community at large.

Compliance is baked into our customs, our laws, and even the way we talk to one another. That isn't necessarily a bad thing. Civil society is largely based on the expectation of compliance. From speed limits to smoking bans, municipal zoning regulations to workplace codes of conduct, ours is a world that runs on compliance.

For any society to thrive, some level of compliance is necessary—it allows us to cooperate with one another. Governments, formal institutions, and laws may allow us to live in harmony despite our differences. Codes of compliance and obedience help us work together.

But unwavering compliance can also have devastating consequences: for individuals, who can lose their sense of autonomy, independence, and self; for governments, which can devolve into paranoid authoritarianism and persecute their citizens, perpetuating inequalities, poverty, and injustice; and for humanity in general, which requires a fine balance between freedom and limitations. A society without any compliance is anarchy; a society with total compliance is fascism.

The lessons many of us learned as children are extremely powerful, and they extend far beyond our individual lives. *Compliance = good, defiance = bad* doesn't just keep children in line. It shapes the world we live in: our laws, our workplaces, our homes.

It also shapes *us* as individuals. Our training in compliance affects what our brains look like. It affects our wiring. It affects how we think.

So what happens when we want to think differently? When we *need* to?

I've spent much of my life working to answer these questions—not just for the field of psychology, but for myself.

Growing Up Good

When I was a child, my father told me that my name *Sunita* in Sanskrit means "good." The *Dictionary of Sanskrit Names* says Sunita represents "she who has good conduct or behavior."

Although I have spent decades studying why people defy, I was known for being an obedient daughter and student. I did as I was told. I got up when I was told to. I had my hair cut the way my parents insisted. I was what schoolchildren in Yorkshire, England, called "swotty"—preferring my books to other pursuits—but as much as I enjoyed mastering my subjects, and the praise I received for being "good," I always wanted to know the reason why others had an easier time resisting the authority of our teachers, parents, and peers than I did.

I grew up in the heart of the post-industrial north of England in the 1980s. Our small three-bedroom house was crowded: with people, with love, and sometimes with friction. Money was tight, and since we had no extended family nearby, we often felt isolated. My parents must have felt the pressure the most. My mother was dealing with four children on her own every day, all while learning a new language and culture. My dad, who'd earned his PhD in metallurgy, the science and technology of metals and alloys, worked as a lecturer at the University of Bradford—a job for which he was overqualified, underappreciated, and certainly underpaid.

Now that I am an academic myself, I sometimes think about him in those early years: a short man in thin wire-rimmed glasses, his cheeks pitted by smallpox scars, his brown suit jacket barely visible beneath the gray raincoat he wore every day to work. I can see him walking confidently through the halls of a university where very few people looked like him, and where he would have to take indignities with a smile and swallow his hurt. The courage that took, the determination—it inspires me now.

But when I was a child, my father mostly shaped me to be quiet.

In many ways, my father was a traditional one. He was the voice of authority in our household. He worked all day and expected order, quiet, and peace at home. He often seemed exhausted and stressed not only by his workplace but also by the weight of his responsibilities, by the strain of supporting a family of six in an unfamiliar country. My three older siblings and I mostly tried to stay out of his way, and when dad told us what to do, we obeyed.

My dad was strict, and sometimes I felt his rules were too harsh—he once pulled me out of bed in the middle of the night to practice my flute because I hadn't spent the mandatory thirty minutes that day playing through my scales.

His rules were strict for a reason: they were intended to give his children greater prospects than the ones he had managed to obtain by pure force of will. It's a dynamic that is familiar to many first-generation immigrant children, who bear the loving burden of their parents' high and often tough expectations. He and my mother had sacrificed everything for us. Specifically, for our educations.

My mother's parents died in an accident when she was a baby. As an orphan and as a girl in India, she received little investment into her education and often told me how she wished she had had the educational experience I was getting—all for free—in the United Kingdom. To my parents, education was more than exams and degrees. It was freedom, it was respect, and it was a ticket to self-sufficiency.

My compliance and obedience weren't only due to my father's rules. As the child of Indian immigrants, I was well-acquainted with the risks of standing out. I knew that we were different, and that we had landed in a place that did not always welcome us with open arms. I heard the racial slurs and registered the vulgar gestures from older schoolchildren; I saw the stares of other shoppers when we walked into a department store; I felt the jostling and pushing directed our way from adults in the street—and though it was of course impossible, all I wanted was to blend in.

When I was nine, my dad bought me leg warmers in bright red—an unexpected extravagance in our frugal household—and although I appreciated the gesture, I wanted to have them exchanged for a less noticeable black. And when that was impossible, I spent an excruciating day at school trying to hide my legs with my coat.

My parents wanted me to be "good"—to obey and not question authority. But they also wanted me to be independent and autonomous, and my father would fight like hell to give me the opportunities that my mother had never had as a girl in India—the opportunities that he himself had clawed and scraped for, both in India and in the U.K..

Sometimes, this felt like a contradiction. Especially since the English education system itself was not conducive to independence or defiance of any kind. My middle school, for instance, accepted corporal punishment, at least implicitly. I once witnessed a teacher beating a boy for bringing him something in a dirty plastic cup. As the blows rained down, the boy was so fixated on his supposed "mistake" that he cried out in terror—not because of an extra hard punch, but because he saw the cup in question fall off the teacher's desk and split open on the floor.

"Oh, no," he gasped, between sobs. "The cup."

In such a punitive educational setting, I learned early on that the way to avoid punishment was to smile and do whatever was asked of you. When that boy was being beaten, I didn't intervene—I watched the thrashing from a safe distance, peeking through the small window in the frame of the door. When I could no longer stomach what I was seeing, I averted my eyes and walked away.

When I saw that teacher a few days later, in the corner shop near my house, I was startled and scared—but he greeted me warmly, unaware that I had seen him beating a student. In my child's understanding, I believed his friendliness was because I was "good," and that if I ever became "bad"—if I was rebellious, or disobedient, or did anything besides what was expected of me—then I would be treated like that boy.

For most of my childhood, I tried desperately to never become the boy in that situation. I studied hard. I got good grades. I didn't talk back to my parents, or get into trouble, or in general do anything to tarnish the image my parents and teachers envisioned for me: that of an intelligent, hard-working, obedient young woman, headed for a career in medicine.

Being "good" was at the root of my name. It was something I had to live up to. And for the most part, I did.

Another Brick in the Wall

When I turned thirteen, my parents sent me to an all-girls Catholic school. The school accepted a proportion of non-Catholics, and my dad felt the quality of the education would be higher there than at the local school I'd been attending.

Like most schools in England, St. Joseph's had a uniform: a blue navy skirt and a checked blue-and-white blouse, with a red sweater over it. "When I look out at you," one of our teachers was fond of saying at the compulsory morning assembly, "I like to gaze out on a sea of red."

That statement made my eyes narrow. During the assemblies, I hummed in my head the Pink Floyd chorus, seeing in my mind's eye the children from its music video, wearing identical school uniforms and marching in lockstep:

All in all, we're just another brick in the wall.

The girl who once objected to standing out in bright red legwarmers now felt uneasy with a teacher insisting we all look the same. Outwardly, I was still compliant. But inwardly, without a doubt, I was changing— and increasingly fascinated by people who were rebellious, defiant, or just didn't fit in.

Two weeks into the fall term, Clara made her first appearance at school. Ten minutes late, with asymmetric but stylishly long brown bangs over her eyes and strikingly dark eye makeup, she strolled into the assembly hall and made her way to a space near the back. She was the only girl in school not wearing the required red sweater. Poised and unbothered by her inability to blend in, she was a whitecap on the headmaster's sea of red, a different colored brick in our school's wall. I liked her immediately.

We became fast—if unlikely—friends. On the surface, we had little in common. I wore sneakers; Clara wore heels. Clara's parents were not strict, like mine—her father was not in the picture, and it appeared to me that her mother allowed her to come and go as she pleased. Though intelligent, with nuanced opinions about politics and culture, Clara didn't really care about school and her studies the way I did.

But we shared one fundamental thing in common: We both stood out. I was one of just a handful of non-white students in the school, and she, though white, was an outsider, too: a new kid, from Hastings, in the south of England, with a different accent than the rest of us. Neither of us knew a single other girl at the school when we started.

Clara seemed like an adult to me, and not just because she smoked cigarettes during breaks between classes. Confident and mature, she appeared to move through the world according to what she—and not anyone else—wanted for herself. I admired Clara's defiant ability to be herself, no matter what anyone said. She wasn't rebellious just to be rebellious. She just seemed to know who she was.

Could I be more like her? I certainly tried—with predictable results.

Inspired by Clara's running commentary on how great it felt to light up the first in a pack of Benson & Hedges, I smuggled a couple of cigarettes she gave me into my school bag. At home, I stole a match from the kitchen and closed the door to my room. I put on my tape of "Close to Me," by The Cure, my favorite band. Then, opening my bedroom window and leaning against my desk, where my algebra homework lay already completed, I lit up and inhaled. It was ashy and hot—nothing like the glamorous experience I'd imagined—but I took another tentative drag, trying hard to inhabit this new, cool, and what I believed to be a more mature persona.

I felt rebellious for about five seconds, the time it took for the smell of the smoke to reach the rest of the inhabitants of our small, semi-detached house.

"*What* is that awful stench?" I heard my brother say from downstairs. "Is something on fire?"

Within a few minutes, I could hear the rest of my family downstairs questioning the smell. Was someone *smoking*?

Panicked, my hands shaking, I stubbed out the cigarette on the concrete sill of my open window, then dropped it to the ground below. It was now obvious to everyone where the smoke had come from. I could feel their disbelief that the person smoking was *Sunita*. That it was I who had misbehaved. I whose name meant obedient, meant dutiful, meant good.

Shame seared into me, as did the fear of facing my father's wrath. I was too afraid to leave my room. I decided to go to bed early rather than face anyone. I spent the night tossing and turning, unable to sleep.

The next morning, when I went down for breakfast before school, my father was already at the table. Seeing him, my heart raced. How would I be punished? Why had he not rushed into the room to lecture me the night before? Would he do so now?

But my father surprised me. He didn't raise his voice. As he poured the milk onto his cereal, he just quietly stated, "You know, you never used to like the smell of smoke. Remember how you always made us sit on the lower level of the bus?"

When I was a child in England, you could still smoke upstairs on public transportation, and I would always insist we sit on the lower level, where smoking was prohibited.

"I remember," I said.

That was the end of the conversation. My father didn't yell or chastise me, as I feared. That's because he knew he didn't need to. He taught me an important lesson that day, one that I've carried with me throughout my life, and my research:

Defiance isn't about emulating anyone else. It's about acting in alignment with your own values.

Smoking worked for Clara. She knew it was unhealthy, but she did it anyway. Not because she was trying to be like someone else, but because she enjoyed it. She did it for herself.

I had tried to see whether I would enjoy it, too—if I could join her and the other girls in the break between classes, flicking their ash with practiced insouciance.

I had tried it to be like the others. But it just wasn't me.

And for the first time in my life, that morning my father was telling me something subtly different. It wasn't that I had to personify the Sanskrit meaning of my name: *Good*.

I just had to be Sunita.

I had to be myself.

Fighting Our Wiring

My upbringing had given me a master class in compliance, but as I got older, some of the old strictures began to fall away.

At sixteen, I went to sixth-form college in another town for my A levels, and my two years there were starkly different from any education I had previously received—there was no school uniform, no religious assemblies, and no strict rules about when students arrived and departed.

At home, too, my parents unexpectedly became more lenient. My father allowed me to go on vacation to Tenerife with Clara and her older sister. I was suddenly permitted to go out dancing. I saw The Cure three times. The summer before I went to university, I worked as a bartender at a "working man's pub" in a small town near Huddersfield. I also began to frequent a Bradford nightclub next to the university, Tumblers—tumble in, and stumble out, we used to say. (The legal drinking age in England was, and still is, eighteen.)

What's more, my father even drove me and picked me up in the early hours of the morning.

"At least then I know where you are," he'd say, pulling up to the curb. Safety was the condition of my freedom.

My parents were beginning to treat me as an independent adult. They listened to my opinions about politics and music, even if they didn't share them—they particularly disliked The Cure. They let me wear the clothes I preferred, and style my hair in a way I'm sure Robert Smith would have loved but that had my mother running out after me when I left the house, hairbrush in hand, saying I must have forgotten something.

These changes were welcome but also somewhat disorienting. I could express myself, but I still needed my father's permission to stay at my friend's house, attend concerts, work in the pub.

Even my early career was a result of complying with expectations. My parents, my teachers, my community all told me that, despite my ambivalence, medicine was "the best thing I could do." I had the grades; what else would I do? So, at eighteen, I left Yorkshire and headed to the University of Edinburgh to attend medical school, which at most universities in the U.K. then was a combined five-year undergraduate and graduate program.

Suddenly, I had to learn how to navigate a world that was not necessarily bounded by my parents' values. I had internalized so many of the lessons about compliance that my upbringing had taught me, and now, as I matriculated to university, I had to figure out for myself what my own values were, and how I could enact them, in a world with different curfews, dangers, and limits.

It is perhaps unsurprising, then, that during my third year at university, I took a year out of medical school to study psychology. I found myself drawn to reading and understanding studies on obedience, authority, and rebellion. There were many, but perhaps none was more influential than social psychologist Stanley Milgram's.

Milgram conducted his now infamous studies at Yale University in the early 1960s, just as Adolf Eichmann's trial for crimes committed during the Holocaust was drawing to a close. As the son of Eastern European Jewish immigrants, Milgram wanted to know if the constant Nazi refrain, heard over and over during the Nuremberg trials—"I was just following orders"—was a psychological reality outside the context of the regime's genocide.

To test this, he told study participants they were taking part in an experiment on learning and memory. Designated as "teachers," participants were instructed by an "experimenter" to read a series of word pairs aloud to a "learner," who was in a separate room, strapped into what appeared to be an electric chair. The learner's job was to recall the word pairs when prompted. If the learner gave an incorrect answer, the experimenter would instruct the teacher to administer an electric shock to the learner. These shocks would increase in voltage with each incorrect answer, from a relatively harmless fifteen volts to a dangerous 450.

The teachers in this study did not know they were in fact the participants in a study on authority and following orders. They also didn't know that, despite the audible pounding on the wall from the learner in the next room, there was no actual shock being administered. Both the learner and the experimenter were actors. But all the participants knew was that a man in a gray technician's coat was telling them to inflict a shock on someone else.

No matter what the participants said—no matter how they protested or grimaced or threatened to quit—the man would tell them to administer higher and higher voltages with every wrong answer.

Whenever a teacher protested—and some did—the experimenter would use four escalating prompts. They were:

1. *Please continue* or *Please go on.*

2. *The experiment requires that you continue.*

3. *It is absolutely essential that you continue.*

4. *You have no other choice; you must go on.*

If, after the fourth prompt, the teacher still refused, the experiment would end. Otherwise, the shocks would increase in strength until they hit 450 volts—powerful enough to seriously injure or kill a human.

Before the experiment, Milgram asked a group of psychiatrists to predict what percentage of participants would keep on administering punishment past 150 volts, the tenth shock level. The psychiatrists predicted that most teachers would not continue past this point—and that fewer than four percent would keep administering shocks past 300 volts, the point at which the learner would pound on the wall of the room and thereafter offer no further answers.

The psychiatrists predicted that only about one person in a thousand would pull the lever over the label "Danger: Severe Shock" for 450 volts, the maximum allowed.

Not so.

The results of the experiment shocked Milgram himself.

Every single one of his study participants pulled the lever for 150 volts.

Every single one pulled the lever for 300 volts.

And a staggering *65 percent* pulled the lever all the way to the deadly 450 volts.

The first time I read Milgram's study, I was riveted, but I wasn't surprised. Growing up during the Thatcher years, I had watched footage of the miners' strikes on television, each evening news segment bringing images of picketers being beaten by police on horseback. The Yorkshire Ripper was at large, killing thirteen women in and around my hometown of Bradford. I remember seeing their faces on the front page of the newspaper that lay on our living room coffee table and hearing the BBC news admonishing women to stay inside after dark, to not drink in public bars, to not stand out in any noticeable way. Having seen the way authority could manifest as violence, and how disobedience often equated to danger, I understood only too well why the subjects in Milgram's experiment obeyed the experimenter.

But Milgram was not only surprised by his results, he was repulsed. Soon after the experiment, he wrote in a letter to the National Science Foundation:

> In a naïve moment some time ago, I once wondered whether in all of the United States a vicious government could find enough moral imbeciles to meet the personnel requirements of a national system of death camps, of the sort that were maintained in Germany. I am now beginning to think that the full complement could be recruited in New Haven.

It is easy to conclude from the Milgram experiment that our predisposition to obey authority overrides even our most basic moral instincts. We are so trained to comply—to think that *compliance = good, defiance = bad*—that we will do immoral things if someone orders us to. Milgram presented his subjects as either obedient or defiant. A simple binary: Participants either shock the learner, or they don't. The study is memorable because of this stark delineation.

But as I pored over Milgram's study, I was also struck by something else. In his analysis, Stanley Milgram noted "the regular occurrence of nervous laughter." Fourteen out of the original forty subjects laughed and smiled as they were instructed to administer what they believed to be painful electric shocks. Milgram found this "unexpected sign of tension" to be "entirely out of place, even bizarre."

I found it completely relatable.

I recalled how I'd felt witnessing the beating of another student all those years ago—and it was not only a tendency to fear authority and obey. I felt something else, as well: tension, a feeling of unease, a sensation deep in the pit of my stomach.

I hadn't wanted to walk away from my classmate receiving a thrashing. I hadn't wanted to be just a bystander, but staying silent was what I had been "trained" to do, even though it felt wrong.

Milgram noted that many of his subjects seemed nervous, especially when the shocks they believed they were administering were very strong. They sweated, stuttered, bit their lips, groaned, indicating extreme trepidation and anxiety.

These subjects, I realized, were possibly just like me: They *wanted* to say no.

In fact, sometimes they even did, or tried to. Some of the participants who were eventually classified as "obedient" actually attempted to refuse to administer shocks. Sometimes they went silent. Sometimes they hesitated before pushing the button. Sometimes they swore or interrupted the experiment to ask the learner if they were okay.

Sometimes, they even asked for the experiment to stop—but resumed giving shocks once they were told by the experimenter that they had to continue. It seemed so clear to me:

These "obedient" people were trying to defy, but they didn't quite get there.

The people showing physical signs of discomfort, asking the learner if he was okay, and telling the experimenter that they wanted to stop, were all indicating that they were not blindly obedient. Many of them were in fact attempting to resist something they thought was wrong.

How many times had I tried to say no to an unfair request from a teacher, or an overstrict command from my father, only to bury my feelings and do what they wanted anyway? How many times had I swallowed my objections to a racist remark, or followed the will of the crowd, just to avoid a scene or to make others comfortable?

And how many times had I practically scarred the inside of my palms with my fingernails, or nearly bitten a hole in my cheek, instead of acting as I felt was right?

I didn't have the language for it as a university student. But it seemed to me that the subjects of Milgram's experiments were not as simple as they were sometimes described. They may not have been outwardly defiant, but they also were not simply "moral imbeciles," as Milgram described them, who simply followed orders without thinking. They were somewhere in the middle, caught between their wiring for compliance and their undernourished will to defy commands they knew were wrong.

Those weird smiles, those nervous laughs—they never left me, even as I finished my degree in psychology and returned to medical school, just as everyone expected. Even as I made my rounds as a junior doctor, I thought about the sweaty palms of Milgram's participants, pushing the lever for shocks.

In the years since first reading his results, after more than two decades of research in psychology, I have come to the realization that defiance and compliance are not binary, but rather exist on a spectrum. Defiance is not merely one state. Instead, it's a process, encompassing a gradation of understanding, questioning, and action.

The truth is, most of us are just like those people in Milgram's experiments. Like them, we're wired for compliance. We've been trained for it all our lives. If we found ourselves in the laboratory of an Ivy League campus, asked to administer electric shocks, we might hope we would refuse. But the likely reality is that we would be more like those dozens of people who sweated and stammered and nervously smiled, torn between two conflicting values: obeying authority and not harming other people.

And just because you push the button for the shock, that doesn't mean the decision was easy—or that your moral code doesn't oppose it.

Failing to defy authority does not make us monsters. It doesn't even make us totally obedient.

It makes us human, and perfectly positioned to change the way we act.

2

Tension Is Your Strength

I often start my graduate school business classes off with a thought experiment. Imagine you're part of a team of engineers trying to break the land speed record. You and your colleagues have spent years designing, building, and testing *Firefly*, a rocket-propelled vehicle that looks like a giant metallic pencil. At full speed, it will be able to exceed the maximum speed ever recorded on land: 763 miles per hour.

Firefly has attracted enormous media attention, in part because it has competition: A rival team of engineers has produced *Falcon*, a similar vehicle that they claim is even faster. You believe *Firefly* to be superior as strongly as the engineers in the other team believe *Falcon* to be the best. To put the matter to rest about which vehicle is in fact the fastest, both teams will run speed trials together, putting their vehicles in a head-to-head race on the salt flats of Utah. It will be the first ever televised race of this kind.

A place in the record books isn't the only thing at stake. There is also a ten-million-dollar cash prize and a lucrative contract with a top auto company awaiting the winner. Victory would make your career (as well as your team's), bring you considerable fame, and secure the valuable funding your team needs.

The race is tomorrow, and all of your team—technicians, managers, and the driver—are huddled in their trailers to escape the blinding desert sun, waiting with eager anticipation for the early morning race.

Everything is on track but for one problem. On occasion, *Firefly's* rubber fuel lines have leaked during a test run. You think the issue might be due to high levels of humidity, and tomorrow is projected to be an abnormally humid morning in Utah—85 percent. The lines have never been tested in anything above 75 percent humidity.

Just before the final pre-race meeting, you share your concern with the management team.

"The forecasts are saying the humidity will be 85 percent tomorrow," you say. "That's like Florida. We never thought we'd encounter conditions like that here."

The lead manager of the team, who has been on the telephone with the television networks all day coordinating camera placement and access for journalists, looks at you intensely.

"Does that mean anything?"

"Well, we've never tested the fuel lines above 75 percent. So, I don't know."

"You engineers!" the manager utters with annoyance. "I've had the president of NBCUniversal on the phone all afternoon, the governor of Utah is flying down in an hour, and *now* you tell me you don't know if the car works?"

"Give me thirty minutes and I'll check the data," you respond.

You quickly gather and examine all the data from *Firefly's* past test runs in which fuel lines have malfunctioned and document the humidity on those days. You analyze the relationship between humidity and the fuel lines leaking.

There isn't any pattern. *Firefly's* lines have leaked in high humidity, but also in low humidity—the kind that would be normal for Utah this time of year.

You chart the data and bring it to the team managers at the final prelaunch meeting. They are relieved.

"So there's no evidence they'll leak?" the lead manager asks.

"Correct. There is no evidence that the rubber fuel lines will malfunction," you say. You hesitate, then continue, "But there's also no evidence they *won't*."

The lead manager sighs. "Let's put it to a vote. I say we go for it."

The decision to race is based on team consensus. But you know that if anyone votes no, the race will be canceled, *Falcon* will win, and you and your colleagues will lose out on millions of dollars, a lucrative contract, and a place in the record books.

You consider your vote. You don't have any conclusive evidence of a problem. But you are tense. The problem of fuel leakage must be fixed, otherwise it could cause an explosion.

"Who votes to proceed?" asks the lead manager.

One by one, hands go up. The whole team is in agreement and wants to go ahead with the race in the morning.

You're the only one who hasn't yet voted. Your heart is beating fast. Sweat beads on your brow. Your hand is at mid-chest, seemingly floating between yes and no.

What do you do?

Ready to Race?

If you said go ahead with the race . . .

I'm very sorry to inform you that you just voted to launch the ill-fated *Challenger* space shuttle.

This scenario is modeled on the situation engineers faced at Morton Thiokol, the aerospace firm NASA contracted to help build the rocket boosters for its space shuttle program. In the waning months of 1985, five Morton Thiokol engineers had concerns that there might be issues with the large rubber seals—called O-rings on account of their shape—on the *Challenger*'s solid rocket boosters. They were uncertain about the O-rings' performance in unseasonably cold weather. When the craft's January launch date forecasted freezing overnight temperatures, they grew concerned enough that they officially recommended that NASA postpone the launch.

The Thiokol engineers had no empirical evidence that the O-rings would function well at temperatures below fifty-three degrees Fahrenheit. But the problem was that they had no evidence, or at least they thought they had no evidence, that the O-rings would *fail*, either. They had examined temperature data from launches in which O-rings had failed, but the results were inconclusive.*

So NASA, for a variety of reasons—perhaps the massive publicity occasioned by the crew, which featured the first teacher in space, or the timing of the launch vis-à-vis President Reagan's State of the Union address—put pressure on Morton Thiokol to reverse its recommendation and approve the launch.

They did, and the rest is history. All seven crew members died that day, in one of the most devastating disasters in space exploration history.

In the years since, we have learned a lot about the decision-making process that launched the doomed space shuttle. And what stands out to me about the engineers' statements is how much they talk about the *tension* they felt about the decision to launch. Over and over in their accounts, they talk about their distress, their dread, their growing apprehension as the launch date drew near. The engineers felt immense pressure to approve the launch. They knew that NASA did not want to hear a "no go," and felt torn between their concerns about the launch's safety and the clear pressure from their superiors to suppress their concerns.

---

*Later modeling of O-ring performance that examined all launches—when the O-ring functioned *well* and when the O-ring *failed*—clearly revealed that colder temperatures increased the risk of O-ring failure.

As a result, when NASA pressured the engineers to reverse their ruling about the launch's safety—and when higher-ups at Morton Thiokol, on a separate call, capitulated to approve the launch—it made them viscerally uncomfortable. Angry. Upset.

"How the hell can you ignore this?" Roger Boisjoly, the lead engineer on the task force investigating the effects of cold weather on the O-rings, yelled on a NASA conference call the night before launch.

Bob Ebeling, the first of the engineers to raise concerns, was so certain of disaster that on the morning of the launch, he drove to Morton Thiokol's Utah offices distraught and slamming his fists on the dashboard.

"The *Challenger's* going to blow up," his daughter recalled him saying. "Everyone's going to die."

Allan McDonald, director of Thiokol's Space Shuttle Rocket Motor Project at the time of the disaster, would later testify before Congress about his concerns that the launch was not safe and should be postponed.

These engineers felt that something wasn't right. And NASA's pressure to reverse their no-go recommendation intensified their discomfort. What they felt was a substantial unease: tension between what they believed to be safe (not to launch) and what they knew NASA wanted (to launch). And despite the lack of empirical data at hand—and at considerable risk to their careers—they didn't try to sweep that tension under the rug or ignore it. They articulated it.

After the *Challenger* disaster—and after testifying before Congress about his efforts to stop the launch—Roger Boisjoly found himself essentially blacklisted and soon resigned. Bob Ebeling retired. Only Allan McDonald remained at Morton Thiokol, where he became a fierce advocate for ethical decision-making. Although these men weren't successful in convincing NASA to postpone the launch—and wished, to the end of their lives, that they had done more to stop it—they never regretted speaking up.

Usually it is only one person, or a small handful of people, like the engineers at Morton Thiokol, who speak up. More often, we humans tend to go along with others in situations like this one. When I share a case similar to the opening scenario of this chapter, almost every single team of students in my classes votes to race. Sometimes, every member of the group votes yes. Often, however, there are initial dissenters, who are either persuaded to say yes or outvoted.

When I was a student, I likely would have gone along with the majority. Back then, in becoming a physician, I had followed the path that had been set out for me since childhood. By going to business school, I took a new direction—but in so many ways, I was still complying with the vision and expectations others had for me. It would take a few more strides before I'd make my own way to the research I felt so driven to explore. Another couple of steps before I would stand in front of a classroom myself.

Back then, if I had encountered the story of *Firefly* and *Falcon*, I probably would have thought that it was too unsafe to go ahead with the race. I probably would have felt the same tension that some of my students express about their decision to race.

But I, like them, wouldn't have known what to do with it.

Inside the Decision-Making Process

"Trust your gut."

"Go with your gut."

"What's your gut reaction?"

As a psychologist, I've always worried about these phrases. They imply an unthinking reliance on instinctive reactions, a kind of "first thought, best thought" approach to life that doesn't leave room for more nuanced reflection. And while the microbiome of our large intestine is indeed an important part of our physiology, we sometimes overestimate how much it should affect our decisions. An instinctive "gut feeling" often comes from past experiences, which are sometimes relevant to the current context, and sometimes wholly irrelevant.

In the case of the *Challenger,* Allan McDonald mentioned before Congress his "gut feeling" that something was wrong about the launch. But it would be a mistake to consign his unease to the same emotional sphere as that of, say, a nervous flyer before a transatlantic flight. His gut feeling was most likely the product of what psychologists call *expert intuition*: an accelerated cognitive process, borne of a high level of experience and knowledge, many thousands of repetitions, and a stable environment that provides immediate feedback. McDonald and the other engineers had spent years developing and studying the physics governing their rocket. If they felt something was "off," that feeling was in many ways the product of earlier deep thinking, practice, and experience. When your "gut" speaks to you, it could be expert intuition—but it could also be your biases. Being able to distinguish between the two is critical.

Expert intuition needs at least three conditions to develop and be accurate: a predictable environment, immediate and unequivocal feedback, and repeated practice. A chess Grandmaster seeming to predict their opponent's next move is often working on expert intuition: they have experienced, many times over, the same situation. They can predict the consequences of each move or pattern of moves. And they have been practicing for years—sometimes most of their lives. When they have a "gut feeling" about what to do, it's their expert intuition informed by years of experience, thinking, planning, practicing, and recognition.

But most times, the conditions for expert intuition are not present. More often than not, in situations that call for defiance, we are faced with a new choice, one we haven't had to make before. We cannot blindly trust our gut. We can listen to it—for it may signal important information—but we also need to qualify it and know when to override it. Allan McDonald and the other engineers may certainly have been drawing on their expert intuition to inform their concerns about the safety of the rocket. But they also had to face a new situation. And what interests me about the Thiokol engineers isn't their expert intuition about the O-rings so much as the *tension* they experienced to keep silent about their concerns.

The tension here isn't about the Thiokol engineers' intuitive understanding that the launch wasn't a good idea, based on years of experience with O-rings, although that is important. This tension came from the engineers' desire to vocalize their opinion, and to persist in vocalizing it, in an environment where they were expected to be quiet. It's about the tension of being caught between those two opposing forces: between what the engineers wanted to do (what they knew was right) and NASA's pressure for them to succumb, to drop their concerns, to comply with NASA's desire to launch. It's the same tension I experienced when I walked away from watching my classmate being beaten, the same tension the Milgram participants appeared to display when they were asked to continue delivering electric shocks to another human. It's likely the same tension you have felt when you want to speak up but find yourself tongue-tied.

Maybe someone at your office has told a joke at someone's expense. Everyone is laughing, but you feel as if you need to say something—you just don't know exactly what.

Maybe you witness sexual harassment on the subway. A man is saying lewd things to the woman next to him, and although she tries to ignore him, his voice sticks in your ear.

Maybe your boss has told you to cut a significant corner. What she's suggesting could be illegal, and although she assures you no one will ever find out, you're uneasy.

This is the tension that arises when we're caught between opposing forces, between our wiring to comply and our desire to act in alignment with our values. And it's a tension we often resist. It doesn't feel good. It doesn't feel clear. And when we're not aware of what it indicates, we try to discard it.

Why We Resist Resisting

When I use the case study that opened this chapter—to race or not to race—my students don't immediately vote to approve going forward. Instead, they pose questions to their team members that show doubt: Does the data really show what we think it shows? Is inconclusive data enough to veto the benefits of a successful race?

Tension often manifests as *doubt*. Perhaps this is how some of the Milgram electric-shock experiment subjects felt, asking themselves: *Is this really okay? Is this safe? Should I really be doing this?*

Tension can also manifest as *anxiety*. Milgram saw this: subjects nervously laughing, sweating, and muttering. The Thiokol engineers certainly lost sleep over their decision to recommend postponing the *Challenger* launch.

But neither anxiety nor doubt wholly capture what we experience in those moments when we're deciding whether or not to defy.

You can feel anxiety about defiance, for example, but have no doubts that it is necessary.

You can be consumed with doubt about how to behave in a situation, without that doubt manifesting itself as anxiety.

That's because doubt and anxiety are symptoms of something more general: tension stemming from an ingrained *resistance to resistance.*

Because we are taught to obey from a very young age, when defiance becomes a possibility, we actively resist it. We feel a tension between what we are asked to do and what we actually want to do. Between what we are told is right and what we know to be right.

That resistance to resistance might feel like a punch in the gut, a flutter in the chest, or a nervous stammer on our lips. It might manifest as cognitive tension, something that arises from our conscience and makes us uncomfortable. The feeling can be extremely useful, but because it so often feels like anxiety, doubt, or discomfort, we may be more likely to see it as a hindrance: a sign of weakness or a lack of confidence.

So when we have doubts about a situation, we often sweep them under the rug in favor of authoritative certainty. We push our tension down so we can avoid conflict, make our lives a bit easier, and avoid ridicule or judgment. We go along with things just to be polite, to appease the offender, and to not create a scene. We convince ourselves that others probably know more.

So we vote along with everyone else in the work meeting to approve a new budget, even though we are unsure: *If everyone else thinks this is the way to go, it's probably right.* We get in the car after leaving the bar with the rest of the group, even though the driver has had a couple drinks: *He's probably fine.*

When we are anxious about what we are being asked to do, we often disregard our tension, assuming there must be something wrong with *us.*

It's easy to rationalize that there is little we can do about something, or that it's not "worth" our tension, our doubt, our anxiety.

And in doing so, we ignore one of our brain's most potent tools for making decisions about defiance.

Use Your Tension

Jeffrey Wigand thought he could do it.

As a scientist with a PhD in biochemistry who had spent decades working in healthcare and biotechnology, he wasn't exactly positioned to become the next vice president of research and development at a tobacco company. But at a crossroads in his career, he needed a job, and he'd heard rumors that the company was attempting to create a safer cigarette. Maybe, he thought, he could work within the tobacco industry to make it safer and healthier. Maybe he could view his work as another healthcare challenge, straddling the line between company man and independent researcher. He felt some discomfort with his decision to take the job, but with some mental gymnastics and the six-figure paycheck, the largest in his career, he thought he could overcome it.

At first, everything went well. Wigand and his family moved back to Louisville, Kentucky, his wife's hometown. They bought a beautiful, large house and put their daughters in excellent private schools. Life was a whirlwind of company dinners, weekend golf outings, and charity events. For a man who had jumped on and off the corporate roller coaster for much of the previous decade, uprooting his family and their finances every few years, it seemed, at long last, that the Wigands had found stability and comfort.

But within a few months, Wigand grew disenchanted.

The lab he inherited at Brown & Williamson was in shambles, more like an outdated high school chemistry classroom than a major research facility. Brown & Williamson didn't even have a toxicologist or a physicist on staff, necessities for studying both the chemical safety of tobacco products and their flammable properties. Very quickly Wigand began to realize that the issue wasn't merely benign neglect. Instead, his employer was actively trying to hamper his research. References to carcinogens and harmful additives in cigarettes began to disappear from the company minutes of research meetings. Important research into the possible harmful side effects of smoking had a habit of disappearing into overseas labs, never to be seen again. Wigand's initial discomfort with the job started to grow.

Wigand began to feel that it was impossible to perform the research he thought he'd been hired to do. He began to fear that even if he could do that research, it would be ignored. He was right. Wigand said later that when he went to Thomas Sandefur, the president of Brown & Williamson, he was flatly told to stop trying to develop a safer product. Doing so would open the company up to enormous scrutiny: If they made a "safer" cigarette, then they had to admit that their other products were dangerous.

So Wigand pivoted. Instead of developing a safer cigarette, he began to investigate the role various chemical additives played in the company's tobacco. Chief among them were glycerol, which kept the tobacco moist during packaging, and coumarin, which helped give some of the tobaccos their sweet taste. Glycerol was usually harmless, but when burned in a cigarette, its chemical composition changed in such a way that it could act like a carcinogen. And coumarin, he learned, caused cancerous tumors in the livers of lab mice.

When Wigand brought these grave concerns to his superiors, he was not praised. He was, instead, reprimanded. His company reviews began to include phrases such as "a difficulty in communication," and his supervisors tried to paint him as too forthright with his opinions. Eventually, after four years at the company, he was fired—and then, after complaining about his severance package to a former colleague, Wigand was sued on a pretext of breaching his confidentiality agreement. Desperate to retain medical coverage that one of his daughters, who suffered from spina bifida, needed, he was coerced to settle by signing an even stricter and longer confidentiality agreement.

Wigand was furious about being fired. More than that, he was astounded by the depth of the deception that the tobacco industry had engaged in—not only with him, but with the American public. For decades, its CEOs had claimed that nicotine was not addictive, that cigarette smoking did not cause cancer, and that their products could not be to blame for the myriad health problems—heart disease, emphysema, etc.—afflicting millions of smokers per year.

Wigand had no doubts: He knew that these were lies. And more importantly, he knew that his company *intended* to mislead the public about the safety of their products. His firing—and the explicit efforts to silence him—was proof.

Wigand wanted the public to know what he knew. But he felt great tension between exposing Brown & Williamson and the immense consequences to him and his family for speaking up. He had already lost his job, and he was terrified to lose the medical coverage, conditional on signing the confidentiality agreement. When a producer from CBS's *60 Minutes* approached him about what he knew, he felt a duty to speak but worried about going on the record. For about two years, Wigand hesitated, torn between his desire to tell the truth and fear for his and his family's safety. Wigand's discomfort did not dissipate with silence, however. His tension only grew.

What happened next was the basis for one of the most important whistleblowing actions of the twentieth century—as well as the plot of Michael Mann's 1999 film *The Insider*. Wigand initially became an anonymous consultant for CBS, helping to decode a trove of documents a paralegal had leaked from Philip Morris, another massive cigarette company. Then he gave a deposition to the attorney general of Mississippi, as part of a lawsuit filed by three states against Big Tobacco. Meanwhile, the producers of CBS's *60 Minutes* were trying to convince him to agree to that interview on prime-time television, telling the country exactly what he knew.

His former employer, Brown & Williamson, caught wind of these plans, and launched a high-pressure campaign to keep him silent about his experiences at the company. They hired a PR firm to dig up dirt on Wigand and spread it in the press.

Wigand began to receive death threats. He found a bullet in his mailbox. Threatening phone calls to his house made ominous mention of his kids' safety. His marriage unraveled. His family was in shambles. He started drinking heavily.

But in the end—after legal battles and months of anticipation—millions of people across the country watched Jeffrey Wigand speak to Mike Wallace on *60 Minutes*. It was one of the landmark interviews in the history of TV journalism, the first time a former high-ranking figure in Big Tobacco went on the record to detail how dangerous cigarettes could be. After Wigand spoke, the tobacco companies could no longer claim that cigarettes didn't cause cancer, or that nicotine wasn't addictive, or that the additives in tobacco would not have adverse effects on people's health. They also couldn't claim that they'd been unaware of these facts for decades.

Wigand became known as one of the most effective whistleblowers in American history. His testimony was one of the watershed moments in the campaign to curb cigarette smoking. It had enormous effects on public health. In the end, forty-six states would sue the largest American tobacco companies, and in 1998 they reached a historic settlement that required Big Tobacco to pay them more than $206 billion over the next twenty-five years and heavily restricted tobacco advertising, marketing, and promotions.

Wigand almost single-handedly brought the tobacco industry to its knees. In the years after his interview, the decades-long decline in cigarette smoking would hasten precipitously.

Jeffrey Wigand undoubtedly saved many lives, but the consequences for him were enormous. His whistleblowing cost him his job, his marriage, and his privacy. The smear campaigns from Big Tobacco put his personal shortcomings—drinking, anger issues, a strained relationship with his family—on full display for the nation to see. The negative effects on his mental health were considerable.

But ten years later, when Mike Wallace interviewed Jeffrey Wigand on *60 Minutes* for a second time, Wigand seemed at peace:

"I don't think I've been this happy in a long time, Mike," he said. "I got a little fatter, a little grayer, but every day I know I've done something that makes a difference for another human being."

When we see tension for what it is—the symptomatic expression of our deeply coded resistance to resistance—then we can use it as a tool. Inner tension is your brain and body telling you that defiance may be necessary. But that does not make it easy. Going against the expectations others have of you often has consequences.

If you act on the tension, it might mean that you look irrational to others, or even to yourself.

It might make you look "emotional" or "difficult."

It might cost you your job, or a relationship.

But whether you experience it as a sinking feeling, a gut punch, your conscience speaking to you, or a knot in the pit of your stomach—whether it feels like anxiety, doubt, or even physical discomfort—*pay attention to your feelings of tension.*

Listening to your tension might also make your office safer for your co-workers.

It could protect that woman on the subway from a serious assault.

It could keep your company out of legal trouble—and you and your boss out of jail.

And beyond the immediate realm of your decision, your tension could inspire greater change: a less racist society, public spaces that are safer for women, and a business world more committed to ethical dealing.

Tension is the signal our brain gives us when we want to defy, even *before* we know we want to defy. Our nervous system is sounding the alarm that something is not quite right, that compliance would be a betrayal of our true ideals, of the person we truly want to be.

And it is when our tension is the strongest that defiance might be most necessary. The truth is, when we feel torn between complying with others and defying them, our resistance to resistance *is* a sign that something is wrong.

But it isn't something wrong with us: it's something wrong with the situation.

When my doctor recommended the unnecessary CT scan, I felt uncomfortable and tense. I was torn between what I wanted to do and what the doctor wanted me to do. I felt the first stirrings of defiance—that queasy, uncomfortable feeling. I acknowledged it and tried to vocalize it in questions about the amount of radiation, hoping the technician would ask me if I wanted to proceed. But when that didn't work, I acquiesced to what was expected of me.

I attempted to ignore my tension, but something strange happened. As the machine collected images of the blood vessels in my lungs that I knew would reveal nothing, I only grew more uncomfortable. The tension didn't dissipate—it just made me feel worse for yielding to the pressure.

That incident taught me not to ignore my tension but to listen to it.

Tension signifies your agency, your choice, your power. If you felt no tension, you would be comfortable simply complying. It is your strength against the tyranny of unexamined compliance, a warning designed to help you notice what matters to you, to focus your attention so you can live in accordance with your values.

Your tension might *feel* like personal weakness, but it isn't. It's a potentially world-changing power. Even one person's defiance can avert catastrophe or help millions—as Jeffrey Wigand's did.

His defiance started with tension—and it was there from the very beginning of his employment at Brown & Williamson. It was there before he even took the job. The scientist in him knew that working for Big Tobacco would put him in a compromising position. There was a fundamental tension between his identity as a man of science and his job as an employee at a tobacco company, between his values and the goals of the organization.

That tension was the start of a process of defiance.

For years, he tried to suppress his discomfort. He wrestled with it.

"I used to come home tied in a knot," he told *Vanity Fair*. "My kids would say to me, 'Hey, Daddy, do you kill people?' I didn't like some of the things I saw. I felt uncomfortable. I felt dirty."

What Wigand describes, in vulnerable moments like this, is the part of defiance that most people learn to hide. The feelings of weakness, of shame, of hesitation, that precede the moment of refusal, of stand-taking, of heroism.

I felt uncomfortable. I felt dirty.

Ultimately, Dr. Wigand didn't dismiss his tension.

He felt it. He listened to it.

Then he used it.

3

Know Your True Yes

April 9, 2004, began like any normal Friday afternoon at McDonald's in Mount Washington, Kentucky: fryers hissing, burgers sizzling, the steady rhythm of cash registers opening and slamming shut. There was no indication that the lives of several of its employees were about to change forever.

Assistant manager Donna Summers was in the restaurant's cluttered back office. In her early fifties, with graying hair cut to her shoulders and thin wire-framed glasses, she had recently become engaged to a man ten years her junior named Walter Nix.

At around five P.M., the phone rang. A deep male voice identified himself as a police officer. He informed Summers that a theft had been reported in the restaurant and told her that the suspect was an employee. "Officer Scott" described a teenage girl, about ninety pounds, with dark brown hair.

Summers thought the description fit Louise Ogborn, but it was hard for her to believe that Ogborn would steal. A responsible, college-bound high school senior who took extra shifts to help support her sick mother, Ogborn had no history of dishonesty, theft, or misbehavior.

But Officer Scott said that she'd taken a customer's purse, and that she also was being investigated for dealing drugs. He said that "corporate" was involved and that he had the store's manager, Lisa Siddons, on the other line. Referring to specific police codes and state laws, he explained that the situation was serious. He was sending officers to the restaurant as they spoke, but he didn't want to give the suspect time to escape or destroy the evidence. Summers would have to help him detain her.

So she did. She called Louise Ogborn into the store's back office—a crowded, windowless space barely large enough for a desk and a chair—and locked the door.

And for the next three hours, under the harsh fluorescent lights—and the watchful gaze of a security camera—Summers did whatever the man on the phone asked her to do.

First, Officer Scott said that a strip search would have to be conducted, so they could make sure Ogborn was not hiding any contraband on her.

Summers obeyed, telling her employee to remove all of her clothing and hand it to her. She shook each item before folding it neatly and placing it in a plastic bag. Even though it was cold in the office, all Ogborn could use to cover herself was her own black employee apron, the one she would normally have been wearing behind the cash register.

When Ogborn began to cry, Summers comforted her, hugging her head to her chest. The strip search seemed over the top to her, but until the police arrived, she needed to watch Ogborn as Officer Scott instructed.

But Summers also had to get back to work as the evening dinner rush was starting. Officer Scott asked her to find someone else to take over surveillance.

Donna Summers called a cook, twenty-seven-year-old Jason Bradley, into the office. When Bradley saw Ogborn, cowering and crying in the back office, he was shocked. He spoke to Officer Scott on the cordless office phone briefly, pacing uneasily up and down the tiled floor. When Officer Scott told him to ask Ogborn to drop the apron, so Bradley could see if she was hiding anything, he repeated the command out loud so that Officer Scott and Ogborn could both hear it—but at the same time, he made eye contact with the terrified teenager and shook his head, indicating that he didn't want her to actually do it.

Bradley handed the phone back to Summers.

"This is a lot of B.S.," he said angrily.

"You can't tell anyone what's happening," Summers said, parroting what Officer Scott had told her, as Bradley walked back to his fryer.

The evening rush was now in full swing, the restaurant filling up with hungry patrons. The line snaked toward the door, and customers were becoming impatient. Summers couldn't spare any other workers to watch Ogborn, she told Officer Scott.

"Are you married?" Officer Scott asked.

"I'm engaged," Summers answered. Despite everything that had gone on that afternoon, she began to giggle in delight.

"Do you trust him?"

"Yes," she said.

"Call your fiancé."

Walter Nix didn't even work for McDonald's. A forty-two-year-old divorced father of two, Nix was an exterminator more familiar with eliminating vermin and insects from buildings than slinging burgers and fries. Burly and balding, with a pencil-thin goatee, he was widely known as a friendly, unthreatening person who coached youth baseball and attended church regularly.

Nevertheless, for the next two and a half hours, Nix didn't just watch Louise Ogborn. Obeying the instructions of the man on the phone, he subjected her to an increasingly horrifying array of psychological, physical, and sexual abuse.

In the restaurant, Summers, Bradley, and the rest of the employees continued to drop fries into oil, stuff paper sacks with steaming cheeseburgers, and take orders for the nightly dinner rush. None of them were aware of what Nix was doing to Ogborn in the back room, away from the routine clamor of a fast-food franchise.

When Officer Scott finally told Nix that he could leave, he got into his car and immediately phoned a friend.

His voice breaking, he said, "I have done something terribly bad."

As implausible as this situation may sound, between 1994 and 2004, the man claiming to be Officer Scott—who sometimes used other names or identified himself as a corporate manager—pulled off a similar hoax more than seventy times, at businesses across the country. It sometimes took hours for the deceived managers and employees to realize that the man on the other line was not who he claimed to be. But until that happened, managers went along with his escalating demands.

They obeyed the "officer" in a Burger King in Fargo, North Dakota. They followed orders at an Applebee's in Davenport, Iowa. They complied at a Taco Bell in Fountain Hills, Arizona.

The enduring success of this hoax might just seem like another example of how we are wired to comply, how challenging it can be to push back even when an action goes against our common sense or values—especially when orders come from people in positions of power.

But it isn't just about how hard it is to say no. It's about how our "yes" can be used against us.

Compliance Is Not Consent

When the cop asks to see your ID, you typically hand it over, even if you weren't doing anything wrong. The TSA agent needs to pat you down, so you hold out your arms like a starfish. When the health club you join wants you to complete a waiver, you probably sign it, barely glancing at the fine print (where you relieve them of any responsibility for your potential death and dismemberment). You go with the program—and in many cases, that's fine. But sometimes it can leave you feeling as if you had no choice.

To comply is to "go along." Compliance is reactive and often imposed: by someone else, by a system, by an environment. There doesn't have to be overt pressure—compliance can be something we passively accept, or allow, or are pressured into: a yes that isn't really a yes.

When I conceded to a CT scan I did not need, I simply did what the doctor wanted me to do. But as I lay in the scanner tube that day, hearing the whirring bits of machinery over my head, I certainly did not feel that I was doing something that I had chosen. I complied, but I did not consent.

Consent is radically different from compliance. It is *a thoroughly considered authorization that is an active expression of our deeply held values.* Consent is powerful and comes from inside ourselves. It represents our True Yes.

As a trained physician, I find the medical definition of "informed consent" to be a useful framework to understand this concept. Informed consent aims to ensure that the rights and autonomy of patients are not violated. It requires five elements: *capacity, knowledge, understanding, freedom,* and *authorization.*

Capacity means the awareness, cognitive ability, or competence to make decisions. Broadly, it refers to a person's mental state: Are they aware of their surroundings and the situation they're in? Someone whose mental state is impaired by drugs, alcohol, or illness lacks capacity.

Knowledge refers to the facts of the situation itself. A person who has been lied to lacks knowledge. So does a person from whom vital information—the harmful side effects of an experimental medication, for example, or the fact that the home they're buying lies in a floodplain—has been withheld.

Understanding refers to how a person processes the information they've been given. Without understanding, the receipt of information is a moot point. Just because a banker gives you hundreds of pages of printed material about the mortgage you're taking out, that doesn't mean you understand that the subprime loan you just signed up for has an excessive interest rate that you cannot afford.

Freedom means that your decision is voluntary, and you have the choice to say no. The choice is purely yours, and you have not been coerced or unduly pressured.

Finally, *authorization* requires an explicit active decision *yes* or *no*. In other words, valid consent necessitates an affirmative "yes," not merely passive acquiescence.

True consent, in any situation, must have all five of these elements. If any one of them is absent, then consent is by definition invalid. And if any of them are compromised, then even authorization itself, a vocal "yes," doesn't signal valid consent—because a person who has been denied capacity, knowledge, understanding, or the freedom to say no cannot give a True Yes.

The McDonald's hoax is a case study in the dangers of compliance, and of how authority can be used to rob people of their ability to say no. On that Friday evening at that Kentucky McDonald's, Louise Ogborn was the clear victim of abuse and coercion. She protested, again and again, when she was told to take off her clothes. Tearfully, she pled with her captors, saying that she'd done nothing wrong.

Ogborn was deceived—she lacked knowledge and understanding. She was coerced psychologically and intimidated physically, trapped in a room without her clothing. She did not receive a choice in how she was treated. She did not feel she could say no—to her manager Donna Summers, to Walter Nix, or to Officer Scott.

"I was scared because they were a higher authority to me," she said later. "I was scared for my own safety because I thought I was in trouble with the law."

The situation was different for Summers and her fiancé, Nix. They had more power than Ogborn and were in less danger. But they, too, were deceived by Officer Scott. They, too, lacked knowledge—and without knowledge, there can be no true understanding. Nix realized only too late, after he left, that something was terribly wrong and that he may have been tricked. Only Jason Bradley, the cook, refused to participate with Officer

Scott's demands. Even then, he didn't openly let Officer Scott know he was not complying with his orders. And, although he was furious with his boss and the situation, he still complied with Summers's orders to keep silent and wasn't able to stop the situation from progressing.

Summers and Nix truly believed Officer Scott was a police officer. The man could describe Louise Ogborn. He knew the name of Summers's manager, Lisa Siddons, as well as the names of other McDonald's corporate personnel. His references to the law and police procedure sounded legitimate. Summers even thought that she could hear a police radio in the background of his calls.

Police officers enforce the law. They demand compliance. Summers and Nix did not think they could refuse. So they didn't. They authorized Officer Scott's abuse, but their authorization was based on false premises.

What happened that day to the manager and her employees was undeniably nonconsensual.

What makes it so disturbing is that most of their actions were also undeniably compliant.

Good at Predicting Consent, Bad at Predicting Compliance

If I asked you to predict how you would respond to a man on the phone telling you to conduct a strip search of one of your employees, my guess is you would tell me that you would slam the phone down.

But would you?

Years of psychological research suggest that there is a big difference between what we think we will do in situations like the Milgram experiment, the *Challenger* scenario, or the phone call Donna Summers received that day at McDonald's, and what we *actually* do. It turns out that we are very good at predicting whether or not we will *consent*—but terrible at predicting whether or not we will *comply*.

Compliance does not require the same level of deliberation, nor the five elements of consent, and so we often simply slide into it, without reflection, directed more by our wiring for compliance than any ideal. Or we reluctantly comply, knowing full well that we do not want to, simply because we don't know how to defy.

Two psychologists, Julie Woodzicka and Marianne LaFrance, presented three questions to a group of undergraduate women, each of whom believed they were interviewing for a job as a research assistant for a professor:

Do you have a boyfriend?
Do people find you desirable?
Do you think it's important for women to wear bras to work?

They had already asked another group of women from the same population what they would do if asked the above questions in an interview. These women predicted they would feel angry and reject the question, confront the interviewer, or walk out of the interview. But the study found that when women were the actual targets of these harassing questions in a job interview, not a single one rejected the question or walked out.

People often experience "empathy gaps" when predicting how they, or others, would or should respond to an emotional situation that is different from their current situation. Women in the actual interview said they did not feel anger in the moment, as the other group of women had imagined they would. Instead, they reported feeling afraid and unable to do anything other than smile uncomfortably.

Smiling in a harassing situation might seem counterintuitive, but it makes sense when we consider that not all smiles represent true positive feelings. Although high-power people—people with power or social status, or those who are part of dominant identity groups—are more likely to smile when they feel so inclined, low-power people—people with lower social status, less power, and those who are part of marginalized groups—often feel the obligation to smile regardless of how they feel.

A colleague once referred to my "crocodile smile" that I display when put in situations where I am not comfortable complying but feel I have little choice—such as being expected to perform the role of secretary at meetings *yet again,* or come in to work over the weekend *yet again,* or politely answer my boss's questions about my "exotic race" *yet again.* It's an instinctive small upward turn at the corners of my lips: the shape of a strained smile. My reflexive smile in uneasy situations is an inbuilt response, created by years of social conditioning that taught me the importance of appearing conciliatory, helpful, and accommodating.

The crocodile smile is a survival strategy, one that women, in particular, often default to as a signal of appeasement. It is a product of the way that our compliance has been assumed for millennia, our consent consistently taken for granted. It doesn't necessarily indicate consent, but it often is interpreted that way. It's a shield that says:

I am not a threat.
I will play by your rules.
I concede.
I comply.

Consent is defiance's twin. It is equally powerful and potentially revolutionary. But it is not always treated as such. The five elements of consent—a person's *capacity, knowledge, understanding, freedom,* and *authorization* to make a decision—represent an ideal, one that I have pursued since I first encountered it in medicine. It is also an ideal that, particularly in my medical career, I have rarely seen fulfilled.

So often, patients lack knowledge or understanding of a procedure, or are so intimidated by the doctor or the hospital environment that they do not feel they can ask questions or refuse treatment. And so authorization, although technically granted via signatures on paperwork, becomes an empty gesture. A check of a box on a medical form seems a false form of consent, especially when a person is in a hurry, doesn't understand, can't read the tiny print, or doesn't know what the other options are. As a physician, I found these situations disturbing, but they did not entirely surprise me. As a patient myself, I had complied with a CT scan that I didn't want. And I'd scrawled my signature on plenty of forms after a long list of paragraphs that I didn't have the time or the energy to read properly.

There is a stark difference between the stated elements for informed consent and their actual implementation in the world. Even a system supposedly designed to empower consent can sometimes push people into mere compliance. And policies that claim to protect patients often end up only protecting institutions. Even as a child, I understood that *yes* didn't mean much if you did not feel you could say no.

Perhaps you recall the slogan "No means no." For decades, it was the dominant catchphrase for consent, aimed at ending sexual assault in schools, on campus, and in our homes. The problem, as we've seen in this chapter, is that "no" often doesn't feel possible in situations of coercion and assault.

Starting in the spring of 2013, a succession of women sued Bikram Choudhury, the founder of Bikram Yoga, for sexual assault, battery, harassment, rape, and imprisonment. The women claimed that the influential yogi—who founded an immensely popular practice of "hot yoga," in which participants execute a series of twenty-six poses in a room that often reaches one hundred degrees—used his popular retreats and training sessions to groom, harass, and pressure his female trainees into sex.

Some members of Bikram's yoga community asked why the women Bikram targeted didn't simply leave—why they kept attending his seminars and receiving his training; why, after being assaulted, some of them even kissed him on the head and politely said, "Good night, sir." These are popular questions among those who have only anticipated—not experienced—such situations. Why didn't you object sooner? Why did you wait so long before revealing the abuse? Armchair observers often attack victims with these same lines, holding up their own predicted responses as the standard to which all other responses are compared. But as research shows, our predicted responses are seldom what we actually do in these situations.

To victims of assault, the answer is clear: When you have been robbed of consent, it is very difficult to assert yourself. Once someone else has taken your True Yes away from you, it can be hard to say your True No. The complexity of being betrayed by someone you trust, in a place that should be safe, leads to self-doubt, denial, and rationalizations. You feel violated but are told that you've misinterpreted things.

In situations like this, to tell yourself that you consented, rather than simply complied, means you are fine. You are okay. You have not been violated.

In a 2014 illustrated essay entitled "Trigger Warning: Breakfast," the anonymous author explains why she made her rapist breakfast in the morning. She needed to believe a different story: that she was in a romantic relationship, that the other person was trustworthy and wonderful. She reasoned, "If I could make him breakfast—eggs, bacon, golden-brown toast—I could pretend that it never happened."

She needed to believe the person she had gone on a date with was moral and hadn't truly assaulted her, because the other reality was devastating.

Hell Yes

Consent isn't just important in medicine and in sexual relations. It is also crucial when signing contracts, authorizing unwarranted stops and searches by the police, participating in research, and providing personal information. Consent matters in all of our personal interactions, in the workplace, home, and society.

A friend of mine has a saying that she uses to avoid having to do what some researchers call "non-promotable tasks": tasks that take time away from important work that is recognized and rewarded. Non-promotable tasks might include planning an office holiday party, even though you're a financial analyst; serving on a committee even though it's beyond the normal scope of your job; or even cleaning up the room at the end of a meeting.

There are, of course, many jobs in which this term doesn't apply because there are few if any promotions to be had. Fast-food workers like Louise Ogborn, for instance, have little power and few opportunities to advance, and they work in environments in which total compliance is expected and indeed demanded, inside and outside the job description.

But in workplace environments with real advancement prospects, non-promotable tasks are often expected of some employees and not others. Even though they require time and effort, even though they take a person away from their real work, many people find themselves completing these unpaid, often thankless tasks.

It can be difficult to refuse. But for anyone who is asked to perform a non-promotable task, my friend's advice is:

Don't say *yes* unless it's a *hell yes*.

As simple as this advice is, it's based on an insight that rings very true: *Real consent often feels good.* And a "hell yes" can be a great benchmark to check in with yourself and make sure you're genuinely enthusiastic about the opportunity and not compromising your values to please someone else. However, it is not the only consideration for consent.

Consent still must satisfy the five requirements—capacity, knowledge, understanding, freedom, and authorization. And a truly transformative understanding of consent would recognize that my friend's maxim is a lot easier for some people to follow than others. Women and those from other marginalized groups often hear that we should say "no" more often, as if doing so is easy. But saying no can be difficult. The same people who are most often asked to do non-promotable tasks have often learned that saying no comes with consequences.

Sometimes our "no" inspires anger and punishment. Sometimes our "no" is simply not accepted, because consent wasn't actually what the person wanted. It was the outcome—the compliance—that they wished to see, not consent. The question was actually a command.

And there are many reasons when it makes sense—for economic, physical, or emotional safety—to "consciously comply." Sometimes the consequences of defiance are too great, or the situation too unsafe, that we may not have much choice to determine our True Yes. In those contexts, you may just have to give your crocodile smile and concede—for the moment.

A Different Perspective

Back in that Kentucky McDonald's, Donna Summers wouldn't know the extent to which her fiancé Walter Nix had tortured Louise Ogborn until hours later, when she viewed the security camera tapes. As far as she knew, Nix had simply been keeping watch. After he left the fast-food franchise, Officer Scott, still on the phone, asked Summers to find another man to take over.

If Summers found it odd that almost three hours had passed and the police still hadn't shown up, she didn't say so. If she found it strange that Officer Scott asked only for men to detain Ogborn, she did not push back. Instead, she simply walked back into the restaurant to look for another male employee. The only one she could find was Thomas Simms, the restaurant's grandfatherly fifty-eight-year-old maintenance man, who happened to be at the restaurant on his day off to get some dessert and coffee.

Summers asked Simms to come back to the office with her, then handed him the phone.

Officer Scott told him the same thing that he'd told Jason Bradley and Walter Nix: Ogborn needed to take off her apron so he could see if she was hiding anything.

Thomas Simms was immediately alarmed. As a newcomer to the situation, he had not heard all the reasons for detaining Ogborn and the escalating demands from Officer Scott.

All he saw in the back room of his workplace was an almost naked, terrified girl, young enough to be his granddaughter.

"You keep that apron wrapped around you," he told her.

Simms handed the phone back to Summers.

"Something's not right about this," he said to her. "This is wrong."

Only then did the hoax unravel. Perhaps it was Simms's words that catalyzed her apprehensions, or perhaps it was the lull in customers that allowed her to reflect on the peculiar requests. But suddenly, Donna Summers's mind was filled with questions.

Why hadn't the police arrived? It had been hours, and the closest station was only a mile away.

Why was it necessary for Ogborn to be watched by men?

Why had Lisa Siddons, her manager, signed off on such bizarre treatment of an employee?

Setting down the office telephone, Summers used her cell to call Siddons, whom Officer Scott had claimed to have on the other line.

"What are you talking about?" Siddons said. "I haven't been on the phone with anybody. I've been asleep all afternoon."

Summers's heart sank. She slowly lifted the office phone back to her ear.

The line was dead.

Over a period of three hours, Summers had been an unwitting accomplice to "Officer Scott." It took Thomas Simms one minute to end the entire ordeal. His uncomfortable feeling punctured the compliant atmosphere established by Officer Scott.

"I know it's wrong, and I would never have been in there to start with if I really knew what was going on," Simms later said in court testimony.

Simms saw and understood the situation for what it was much more clearly than Summers. He, much like Bradley, had not been initially recruited by Officer Scott and did not feel the same need to please an officer or help catch a supposed "thief." It was also Simms's day off. He was not in uniform and was doing Summers a favor by leaving his dessert and following her to the office to help with something completely different from his normal job.

In contrast, Summers had already complied, which made it all the harder to start questioning orders now.

Defiance in the moment often comes early—or not at all. It can be difficult to stop complying once you have started. Doing so typically requires a new or different perspective or taking the time for some active deliberation to consider the elements of consent.

Using the Elements of Consent

When asked to do something you are unsure of, the first step is to pause. Do not say *yes* or *no* right away. The pressures to comply are often most powerful when they are most immediate. We've all endured the "hard sell" of the car dealership or the telemarketer.

That's why it is critical, when making a decision that requires your consent, to take some time and space to think. Even a few moments of taking a "break" in the situation—going to the bathroom, answering a call, turning away—can make a difference and give you an opportunity to ask yourself the questions that match up with the five elements of consent:

> Do I have the *capacity* to make this decision?

> Do I have the *knowledge* necessary to make this decision?

> Do I *understand* what is being asked of me, and the consequences if I say yes?

> Do I have the *freedom* to say no?

If your response to all of these questions is *yes*, that's when you know you're ready to truly answer:

> Do I consent to this?

If yes, you can then *authorize* (the fifth element) your valid informed consent—you can give your True Yes. If no, you can give your informed refusal.

Answering these questions—or even attempting to answer them—can change the way you say yes. It can also help change the way people ask for a yes. These changes are necessary and vital to establishing a world in which consent is not taken for granted. A world in which we respect one another's autonomy, even when the voice of authority says otherwise.

Change is not easy. But it begins with understanding individual truths:

My compliance is not to be mistaken for consent.

My consent is not to be taken for granted.

Only I can grant a True Yes.

The difference between compliance and consent is never a mere matter of semantics. It is a critical step on the road toward defiance:

For "no" to really mean no, we need to know when "yes" means yes.

4

Break Free from Influence

A few months into my first job as a physician at the Western General Hospital in Edinburgh, Scotland, I received an invitation to meet with a financial advisor for a free consultation. I remember that meeting well. It seemed rather important and was held in the hospital's posh meeting room (though as part of the U.K.'s budget-conscious National Health Service, it was not as elegant as you might imagine).

As a young doctor with an exhausting clinical rotation, I spent every other night in the hospital's on-call room, on the cold upper floors of the Victorian clock tower building that housed the wards I was responsible for covering. Bare, dark, and underheated, the large room was host to a single narrow metal bed so creaky that you could easily believe it had been there ever since the hospital was founded, as a poorhouse, in 1868. Dead flies littered the dimly lit stairwell, crunching beneath my shoes as I ascended the stairs in the early hours of the morning, after my final round was complete. Most nights, my beeping pager awakened me shivering and stiff from a fitful sleep. After what seemed like only seconds of rest, I'd find myself again traversing the old stone staircase down to the wards below, with just the physical exertion of rushing to attend to an emergency or making my morning rounds to warm me up again.

I was constantly tired, overworked, and underpaid as a junior doctor in my first year post medical degree. I wasn't quite sure what there was to discuss about my finances, but I accepted the meeting out of curiosity. At the appointed time, after my thirty-hour shift was complete, I entered a room I hadn't even known existed at the hospital. I sank deep into a soft, comfortable couch, admiring the feel of the heavy pile blue carpet underneath my shoes, rather than the cold hard flooring of the hospital wards. As I waited for the financial advisor to arrive, I began to relax. I almost fell asleep.

But then the door opened, and a man named Dan walked in. Cheerful, handsome, and dressed in a sharp fashionable suit, he greeted me with a big smile that reminded me of the friendly drug reps who visited us on the wards. Unlike me, he looked well rested and freshly showered.

"Good afternoon!" he said, as I struggled to my aching feet to shake his hand. "Let's talk about your finances and what we can do with them."

"Sure," I answered. "I'm not sure how much there is to say—I'm quite new, and my salary is not particularly high."

"No amount is too small," Dan said, laughing. "Investment is about building for the future."

Dan was charming, with a positive attitude. He took me seriously, asking me many questions about my spending, my salary, and my goals for the future. He listened and built up a nice rapport with me.

After about an hour, which felt like a vast amount of time to discuss my extremely limited disposable income, Dan recommended that I invest in a particular mutual fund. He also offered to write a detailed report for me about how much of my salary I should put aside each month. What's more, he would follow up with me in about a week. All of this was for free.

I was impressed and couldn't believe my luck. Why would he do all this for me for free?

"What's in it for you?" I blurted out.

Dan smiled and leaned back slightly in his chair.

"Of course, there is no such thing as a free lunch . . ." he said. "I'll receive a commission if you invest in the fund I'm recommending."

My eyes widened. I had in fact received many "free lunches" from drug company reps in those early days of my medical career, when I was otherwise mostly subsisting on beans on toast. It struck me as odd, the way that the reps could shower doctors with gifts and fancy dinners. I'd once looked on in amazement as a drug rep ordered an eight-hundred-dollar bottle of dessert wine at a group dinner, which he opened just after starting a short spiel on a new medication. Once, a senior doctor I was working with told me that such gifts were the just rewards for a profession that—at least in the U.K.—didn't get annual bonuses or company cars. That's why, he said, as he stuffed his pockets with the pharmaceutical company–branded pens and notepads that happened to be left sitting on a nearby hospital trolley, he always took advantage when he could.

Although this doctor told me that those pens didn't affect the way he wrote his prescriptions—and although I told myself that the glass of wine didn't impact my own—it seemed impossible that such gifts didn't come with strings attached, that they didn't impact how doctors made decisions about treatments.

Or how financial advisors pushed mutual funds.

For the previous hour, I had thought Dan had been giving me good advice. He seemed like a nice guy. I liked him. The situation had been friendly and normal—relaxing, even. I'd quite enjoyed myself.

But now that his ulterior motive was revealed, I didn't trust him as much. However, I also didn't want him to *know* that I didn't trust him. In fact, I felt *more* pressure to go along with his advice, just to avoid signaling my distrust to him.

Not only that, I also felt some pressure to appear helpful. Now that I knew what he stood to gain if I chose his fund, it seemed rude if I were to deny him that. I would in essence be telling him that I knew he would lose money because of me, and that I cared more about my own finances than his.

Which would have been true, of course! I wasn't meeting with Dan to boost *his* finances, but to improve my own.

And yet as I sat there on the plush couch, my eyes bleary from lack of sleep, a cup of lukewarm tea on the table next to me, it didn't feel that way.

Independence versus Interdependence

What I know now—which I did not know then—is that even when all the conditions for consent appear to be present, we can still get tripped up by some very powerful social forces that play on our emotions and our anxieties about what other people expect of us, decreasing our sense of freedom to choose. Again and again, my research has demonstrated the powerful role of other people in our choices.

In daily life, we often feel conflict between our individual preferences and prioritizing the wishes of others. Although some Western cultures, especially the United States, highlight the importance of being independent and asserting oneself, people still feel the pull of attending to and fitting in with others.

Psychologists Hazel Markus and Shinobu Kitayama described two different ways we can view ourselves. One is the ideal *independent* view, in which we celebrate our uniqueness, our freedom of action, and our agency. The other is our social or interpersonal identity—our ideal *interdependent* view—that views ourselves in relation to others and is motivated to maintain social norms and harmony over acting on internal wishes.

Our independent self wants us to act on our personal beliefs, no matter the context. Our interdependent self tells us to consider our actions' effects on others, to contemplate social consequences.

These two different ideals we hold—for an independent and interdependent self—are often in tension. We are constantly in pursuit of one ideal or the other, and neither ideal is attainable in its fullest—it is an ideal, after all, not a reality that can be reached.

The truth is that we are always connected to others, while also containing within ourselves our own beliefs, preferences, and values. While many external factors make defiance difficult, this internal conflict between our

warring desires—to be true to ourselves, but also to be connected respect-
fully to our social environment—is one of the most basic and powerful
constraints on defiance. It can lead us to take poor advice. It can cause us
to appease people who might take advantage of us. And sometimes it can
make us confused about what we actually want, and how we actually want
to behave.

That day with the financial advisor, I felt torn: between what I wanted
for myself, and how I felt I should behave toward Dan.

I recognized the feeling. I'd felt it before. Often. In fact, there have
been times in my life when I have felt it every day, from morning to night,
slowly eroding my sense of self as I constantly prioritized other people's
interests above my own.

This feeling is one way our resistance to resistance manifests itself. But
it is more specific than that. I experienced this emotion so frequently
throughout my life that I decided to investigate it in my research. I now
call this feeling *insinuation anxiety*: the concern, worry, or apprehension
most of us feel about signaling a negative opinion about another person
to that person. But the name for the feeling came years after I first acknowledged its power.

The Power of Insinuation

Insinuation anxiety is one of the reasons we don't speak up in the back of
a taxicab when we notice the driver taking an obviously circuitous route.
It is one reason why we nod approval to barbers and hairdressers and say,
"Yes, that's great, thank you," even when we've received a mediocre hair-
cut. And it is one of several reasons why we don't speak up about unethical
or illegal interview questions or the inappropriate transgressions of others
at work.

Insinuation anxiety encourages us to act against our values and preferences in order to protect another person's feelings. We do not want to *in-
sinuate* that we think the other person may be biased, corrupt, or plain

incompetent. So we often comply with a suggestion, keep silent, or accept a bad piece of advice, just so that the very person who is hurting us, costing us, or putting us at risk can "save face."

In my meeting with the financial advisor, I did not want Dan to know that I no longer believed he had my best interests at heart—that in actuality, he was more motivated by his commission than my financial well-being. I instinctively privileged the feelings of the other person. This is understandable. We all share an innate desire to connect and maintain harmonious relationships—some of us more than others. This desire for harmony often causes us to avoid embarrassing other people by insinuating that they are something other than what they should be.

My research has shown that insinuation anxiety is extraordinarily powerful and common. It suffuses our relationships with people at work, at home, and even with strangers. You likely have felt insinuation anxiety, even if you didn't know the name for it.

But once you know what to look for, you may recognize it arising in many different situations. Whenever I talk about insinuation anxiety, my colleagues and friends share story after story about their experiences.

A colleague told me about an experience he had when remodeling his apartment. He and his partner, both dedicated home cooks, had designed their dream kitchen. Every inch of counter space, every cabinet and shelf, was meticulously accounted for.

But when they met with the contractor they had hired to complete the renovations, he told my colleague that the sink would have to be moved.

The next time the contractor came to their house, both of them were present, and they were determined to tell the contractor exactly where they wanted their sink. But when they did, and the contractor again insisted the sink would be better where he recommended it, they folded, reluctantly conceding.

"I didn't want to imply that he didn't know what he was doing," my colleague told me. "It was incredibly difficult to disagree with him. Now, every time I see the sink in that inconvenient position, I shudder."

My friend Rick, who was suffering from lower back pain, went for what he thought would be a serene, relaxing massage. The experience was excruciating from start to finish. Each time the massage therapist drove his sharp elbow into Rick's back with muscle-crushing force, Rick stifled his shrieks of pain. He wanted to speak up, to tell the therapist that he was using too much pressure, but something made him hesitate. Even when the massage was finally over, Rick didn't say anything about how painful it had been. He simply hobbled out of the room, paid his bill, and with a subdued nod of approval even tipped the therapist.

When Rick got home, after a painful subway ride, he told his wife what had happened.

"That's not normal," she said. "Why didn't you say anything?"

"I don't know," Rick said. "I felt uncomfortable. I didn't want him to think that I thought he was doing a bad job."

"So you'd rather have him hurt you than hurt his feelings?" she asked, incredulous.

"Exactly," he said.

If insinuation anxiety can keep us from objecting to an unnecessarily painful massage, it's easy to see how it can play a role in other forms of unwanted touch. This particular feeling is powerful and paralyzing, and what's more, it is not limited to actions that affect only our own bodies. Insinuation anxiety can also keep us from speaking up to ensure the health and safety of others.

According to a 2012 survey, it is distressingly common for crew members on commercial airliners not to speak up when they notice that their superiors—pilots, technicians, senior crew—make a mistake. The survey of more than 1,700 crew members showed that they spoke up only half of the time when they noticed an error. A 2005 survey found that only one in ten healthcare workers—many of them nurses—felt comfortable speaking up when they observed a colleague making a mistake or taking a shortcut.

These silences can lead to potentially dangerous incidents. But even when the stakes are not life or death, insinuation anxiety can have a detrimental effect on our workplaces, our relationships, and our health.

At work, where collegiality is expected and important, it can be extremely difficult to go against the flow of popular opinion or to say anything that might imply that your colleagues aren't acting in good faith. In her book *How Professors Think: Inside the Curious World of Academic Judgment,* Harvard sociologist Michèle Lamont quotes a Black female academic, newly appointed to a funding committee, who objected to the way her colleagues were discussing a particular researcher. She felt that they were relying primarily on gossip to make their decision, and she wanted to speak up in the researcher's defense, but found that she couldn't.

"The one thing that I could not do is what I wanted to do . . ." she said. "That is, to just challenge them. . . . But having just met all four of these people for the first time, I didn't want to question their integrity."

She didn't want to imply that her colleagues were unduly biased. She felt pressure to swallow her objections—pressure that was likely only heightened by her status as a woman of color in a room that was usually full of white men.

This experience is common across disciplines, professions, and a wide variety of other settings. No one wants to imply that their friends, their colleagues, or their family members are prejudiced, corrupt, incompetent, or small-minded. More to the point, no one wants these people to *think* that *we* think ill of them. The result is a charade of good intentions, one in which a kind of conditioned politeness often stifles a true expression of what we really believe. In this way, insinuation anxiety helps preserve the status quo.

Doctor's Orders and a Ferry Ride

Insinuation anxiety might be an everyday feeling, but it is a powerful mechanism keeping us from asserting ourselves, speaking up, and rejecting bad advice. Early in my research, my colleagues and I set out to study the effects of disclosures about conflicts of interest, the kind I experienced with my financial advisor at the hospital. Typically, conflict of interest disclosures are intended to increase transparency: between a financial advisor and a client, between a salesperson and a customer, or between a doctor and patient. The idea is that recipients of such information should be able to account for it in their decision-making, for example discounting advice once they know their advisor may have an ulterior motive.

But we were interested in any unintended effects of such policies. How did the disclosures *really* affect how people make decisions? Did others feel the way I did when confronted with Dan's disclosure of his commission? Did transparency make things better or worse for the recipient?

In a series of experiments, we asked people to imagine they were a patient visiting their physician. In every case, the doctor gave them the same advice about what to do. But some of our "patients" also heard a disclosure about a conflict of interest the doctor had: He would receive a fee or some other benefit if the patient did what he advised.

Our results showed that people had less trust in their doctor when he disclosed a conflict of interest. So far, so good—arguably a decrease in trust is the intended purpose of such disclosures.

At the same time, however, they also felt *much* more pressure to do what their doctor recommended they do. The conflict of interest disclosure left patients feeling torn between feeling less trust in their doctor but greater pressure to comply with their advice.

Patients may decide to decline their doctor's recommendations for various reasons—they might worry about the side effects of a new drug, or they may want to stick with a treatment they had in the past. But hearing about their doctor's conflict of interest didn't make it easier or more clarifying for the patients in our experiment to make their decision. It made it *harder*, because it turned the patient's decision about their own health into one that also included worrying about signaling distrust to their doctor. Rejecting advice became tantamount to calling their doctor a crook. Instead of functioning as the warning it was supposed to be, disclosure became a burden on those it was supposed to protect.

We often feel insinuation anxiety when we are uncertain about the quality of someone's advice—especially when that person is an authority figure or is supposed to have expertise. We may not trust them, but because they are experts, we feel pressure to hide our apprehension. In our studies, whether the patients took their doctor's advice or not depended on which force was stronger: the pressure they felt from insinuation anxiety, or the amount of trust they lost in their doctor. In other words, their ideal selves—independent and interdependent—were in conflict. Torn between a desire to act on their own preferences and a reluctance to send a negative evaluation of their doctor's recommendation to their doctor, the patients in our study felt anxious, stuck between a rock and a hard place.

In the end, many of them followed advice they did not trust, simply to avoid indicating distrust to their doctor—an authority figure who was supposed to have their best interests at heart.

Insinuation anxiety when someone is supposed to know better and have your best interests at heart may be understandable. But what about advice from a stranger with no obvious expertise? That should be pretty easy to dismiss, right?

In a study we ran on a Long Island ferry crossing with just over 250 participants, we discovered that the answer to that is still *no*.

A middle-aged white man—dressed in a suit and tie—asked passengers to fill out a short survey in exchange for five dollars. When they did, he then offered two choices: take the crisp five-dollar bill he was holding or enter a mystery cash lottery where they could have the chance at more. The lottery paid out anywhere from zero to ten dollars but, they were warned, it paid out *less* than five dollars on average.

Overwhelmingly, given just the two choices, people preferred the cash—only eight percent of survey-takers picked the mystery lottery. But the passengers' behavior changed when the man in the suit started offering advice. When he simply advised people to choose the lottery, the number of people who did so more than doubled, to 20 percent. And *further,* when the man revealed that he would get a bonus if they chose the lottery, the number who did so more than doubled again, to 42 percent.

When asked, those participants confessed that while they trusted the man less once they knew about his ulterior motive, they felt more pressure to follow his recommendation. They didn't want him—*a complete stranger*—to think they thought his advice was bad or biased.

And, although we didn't predict it, we uncovered a large gender difference: the participants who felt pressure to follow the man's advice were all women. There are many possible reasons for this, and many of them point to the ways social dynamics can change based on who you are. Perhaps women felt more pressure to be compliant and accommodating to the man. Or perhaps they expected more benevolence from their "advisor" than men did, so they felt greater insinuation anxiety to dismiss his advice.

What this tells us is that insinuation anxiety is powerful, but it is often more powerful for some people than others, and in some circumstances over others. If it can be hard to push back on the advice of a random stranger we do not trust, imagine the difficulties we could have with those whom we believe have our best interests at heart: our doctors, our colleagues, our friends, and our family.

Like so many other pressures that bring our independent and interdependent selves into conflict, insinuation anxiety muzzles us and keeps us compliant. But it is not the only powerful psychological process that can turn us against our better judgment.

The Pressure of the Pitch

The *sales pitch effect* is another psychological factor that plays out in situations that are more transactional. When you're buying a car, for example, you know a salesperson's paycheck depends on their ability to sell you one. While insinuation anxiety is common in situations where we expect the other person to place our interests first, like at the doctor's office, the sales pitch effect is present when we expect some amount of self-interest on the part of the advisor. Rather than an unwillingness to signal distrust or a negative evaluation, it is the pressure to avoid appearing unhelpful, uncharitable, or uncooperative. Both insinuation anxiety and the sales pitch effect were present in women who heard their male "advisor" in the ferry study disclose his conflict of interest. Like insinuation anxiety, the sales pitch effect stifles our opposition by convincing us that someone else's feelings and interests are more important than our own.

To further investigate the sales pitch effect, my colleagues and I invited members of the public to take part in an experiment on board a mobile behavioral science laboratory: a large truck we parked in busy locations around the center of Pittsburgh, Pennsylvania.

As a PhD student, I spent hundreds of hours on that truck, running experiments. On weekends, my husband would bring our toddler to visit, and he delighted in the noise of the truck's roaring engine, as well as the sight of the bright colors and logos on its side: green computers, orange stick figures, and the words RESEARCH TO GO painted in large lettering. Several years later, my son saw a photo of this truck, and recognizing it, excitedly said, "Mummy, you used to live there!"

While my son was far too young at that time to participate in our experiments, most people walking by were not. Anyone above the age of eighteen could play for easy prizes in exchange for just a few minutes of their time. Our experiment was a kind of carnival game—but unlike most activities you might see on the midway or boardwalk, the prizes were much easier to win.

From our shiny data truck, we cheerfully offered people passing by the chance to participate in one of two lotteries, which with a notable lack of imagination we named A and B. The two lotteries had different sets of prizes. Once they selected a lottery, all the participants had to do was roll a fair die. Each number corresponded to a particular prize: candy bars, gift cards, or soda.

It was easy to see which lottery was the better choice. Lottery A had prizes worth more than twice the expected value of the prizes in lottery B. When we asked people to choose between them, without giving them any advice, 98 percent preferred lottery A (see figure 1).

While the players were focused on the prizes they could win, I was more interested in how the presence of another person giving them advice—particularly conflicted advice—changed their choice of lottery. To explore this, we randomly selected some participants to play the role of an "Advisor" and others the role of a "Chooser." Advisors had to give Choosers their written recommendation whether to select lottery A or B.

Number	Die A	Die B
⚀	$20 Amazon.com voucher	Milky Way bar
⚁	$5 Options gift card	$5 Barnes and Noble voucher
⚂	Snickers bar	Mr. Goodbar
⚃	Can of Walmart Cola	Can of Coke
⚄	$5 Dunkin Donuts gift card	$5 Starbucks gift card
⚅	Toblerone bar	$5 Gap gift card

Figure 1: An example of the die-roll lotteries

But there was a catch. Advisors were told that they would get a bonus themselves only if their Chooser picked the less valuable lottery B, giving them a clear incentive to recommend its lower-value selection of prizes to the Chooser. The Advisors were a bit like carnival barkers, convincing people to pick their game, but the difference was that half of the Advisors had to tell their Choosers about their conflict of interest by writing out a disclosure statement at the top of their written recommendation. It would be as if the man working the ring toss had to reveal that he would benefit more by steering you toward an identical game with smaller prizes. The other half of the Advisors were instructed not to disclose their conflict of interest.

I was curious to see if people would follow the obviously bad advice from a stranger, and if the disclosure statement made them *more* or *less* likely to take that biased advice.

The results were striking.

When Advisors recommended the less valuable lottery B (which, un-surprisingly, most of them did) without saying anything about their in-centive, approximately half of the Choosers followed the bad advice and chose lottery B. That was bad enough, but when Advisors recommended the inferior lottery B *and* informed the Chooser of their conflict of interest, compliance went sky high: More than 80 percent of Choosers in this condition chose the less valuable option.

This was a shockingly high level of compliance. Why would people be *more* likely to take bad advice when they knew their Advisor had an ulterior motive to give that recommendation?

Although they trusted their Advisor less when they knew about their conflict of interest, Choosers stated that they felt much more uncomfortable about turning down the Advisor's recommendation in front of their Advisor. They felt pressure to ensure their carnival barker did not lose out. So even though the advice was obviously bad, and even though they did not trust the stranger giving it, most people still went along. And this time there was no gender difference; men and women alike succumbed to the sales pitch effect.

To be clear, this effect wasn't warm altruism on the part of the Choosers (even though they may have liked to think it was), nor were they rewarding their Advisors for apparent "honesty." In a follow-up experiment, we gave Choosers an opportunity to change their mind about their selection—alone, without the Advisor present—and the percentage of people who chose the less valuable option dropped precipitously, from 88 percent to 50 percent. This result was one of the most striking in our series of experiments, revealing how the simple physical presence of another human suppresses our real preferences.

We don't only follow advice when it is good. We often follow it when it is bad—even when we *know* it is obviously bad—to avoid appearing unhelpful to the other person.

The feeling I had that day in the hospital meeting room, of not wanting to signal distrust to my financial advisor, is a classic example of insinuation anxiety. But it was magnified by the pressure I felt not to deny him his commission: the sales pitch effect. The two together can be a powerful cocktail that makes us say "yes" when we really don't want to.

So how do we fight it?

5

Reclaim Your Power

A s a young doctor in that fancy hospital meeting room, I dreaded telling the financial advisor that I didn't want to invest in his fund. He was sitting across the table, waiting expectantly, ready to give me forms to sign. What would he think if I rejected his advice?

Luckily, I didn't have to find out. At the exact moment when the silence between us was becoming truly awkward, my pager started beeping.

"I'm sorry," I told him. "I have a patient to see."

Dan looked mildly deflated, but he rallied with a quick smile.

"No problem," he said. "I can mail you all the materials we discussed today, and you can decide what you'd like to do. Here's my card."

We shook hands and parted on good terms, and the second I emerged from the meeting room, my feet leaving the plush blue carpet for the cold sterile tile of the hospital corridor, I felt a sense of relief.

The papers from Dan arrived promptly, three days after our meeting. But that week, as I trudged up and down the hospital corridors, my back aching from another night lying on the creaky metal bed in the on-call room, I realized just how unsuitable it was to invest what little money I had in a mutual fund. And since I didn't have to tell Dan this directly—since I wasn't in the room with him, and didn't have to imply that he was corrupt, or worry about depriving him of a commission—it was easy for me to reject his advice.

I let the papers sit on a table next to my telephone for a few weeks, then I threw them away.

The Power of a Pause

I hadn't intended to resolve the situation with the financial advisor by leaving the room and making my decision later. It simply happened by chance.

But the way it turned out, however inadvert, reveals one simple solution to mitigate the pressures I've described. Taking a pause and keeping our distance—both physical and psychological—is an effective way to decrease the effects of insinuation anxiety, the sales pitch effect, and so many of the other barriers to enacting what our ideal independent self wants to do.

In my medical scenario studies, the "patients" experienced significantly less insinuation anxiety when their doctor's conflict of interest—that he would receive payment if they followed his advice—was relayed indirectly through a note from the "medical clinic director" handed to them by a receptionist. And in the lottery studies, the Choosers felt far less of the sales pitch effect when they could make their decision in private away from their Advisors.

That's because insinuation anxiety and the sales pitch effect rely on the power of social interaction. If that interaction disappears or is weakened, then so does your anxiety about signaling distrust or unhelpfulness. We can't signal something negative if the other person is not present to send a signal to.

If a car salesperson tells you that they don't need you to make a decision today—that you should take the weekend to think about it—you are much more likely to resist the urge to buy a car you can't afford. If, despite a realtor's insistence that you immediately submit a high offer for a house, you take time away from the pressure to think things over, you are less liable to rush into purchasing an overpriced property. And if your new dentist unexpectedly recommends that you upgrade all your fillings and have your wisdom teeth removed, making the decision away from the dentist's chair, perhaps after seeking a second opinion, could prevent costly and unnecessary work on your teeth.

At the hospital, when I was called back to the ward, I no longer had to look at Dan while deciding what to do. My concern about signaling a negative judgment to him mattered far less in his absence. What mattered more were the numbers on my monthly budget. And in the cold gray light of a wet Edinburgh afternoon, it was obvious that I did not need to invest any money with him.

When we feel pressure to comply, the best time to make a decision is *not* in the heat of the moment. To preserve our ability to give a True Yes or a True No, we must minimize the extent to which these social pressures affect our decisions. Saying no to bad advice is easier when we create space between us and the person giving us the advice.

To overcome these social pressures to comply, the distance between you and the other person doesn't always have to be physical. Increasing psychological distance can also work, if exiting the situation is hard. One of the simplest ways to achieve psychological distance is to talk to yourself—even silently—as though you were someone else: a trusted friend, a mentor, or even just another version of yourself.

For example, as a young doctor feeling pressure to invest in front of Dan, I could have closed my eyes and asked myself:

Sunita, does this make sense to you? Do you really want to do this?

Third-person self-talk might sound trivial, and of course it's much easier to do this in private rather than in front of another person, but research shows that talking to ourselves—even for a moment—helps us regulate our emotions so we can make better decisions.

So does simply naming the feeling. When we know that what we're feeling is insinuation anxiety or the sales pitch effect, it is easier to recognize the situation for what it is.

Remembering the elements of informed consent, we can determine if we really feel free to say no. Overcoming insinuation anxiety and the sales pitch effect is not about your level of intelligence or goodness, how courageous you are, or whether you are an extrovert. All of us—regardless of gender, age, or personality—can succumb to these pressures. It's natural to want to maintain harmony in our relationships. But we can get so wrapped up in being polite that we forget who we are.

The discomfort arising from these psychological processes is often our first indication that a situation calls for resistance. That tension is a warning sign. When we feel it, identify it, and name it, we can recognize what is happening: our ideal independent self is in tension with our ideal interdependent self. We can embrace that conflict to see ourselves for who we truly are—neither totally independent, nor totally dependent on our social surroundings.

There is something deeply human in that tension, that simultaneous desire to both be true to ourselves and in tune with others, and something valuable as well: these feelings can also push us to reexamine how our decisions affect others. A society in which everyone acts only according to their individual interests and desires is not one you'd probably enjoy living in. As the saying—often attributed to Supreme Court Justice Oliver Wendell Holmes, Jr.—goes, "The right to swing my fist ends where the other man's nose begins." Some of the pressures I've described, the uncomfortable feelings, are uncomfortable for a reason: They remind us that our decisions have consequences for other people. Sometimes, we may choose to give up something for the good of someone else, or for a larger collective. The point is that you become *aware of it as a choice* and feel empowered and able to say no.

Compliance Is Our Default, but Not Our Destiny

I remember it as if it were yesterday: I was seven years old, and my mother and I were walking home from the supermarket, our rickety shopping cart loaded with groceries. Our usual path home was through a snicket—that's what we call an alley in West Yorkshire. That day, we were confronted by a group of five or six teenagers, the kind of boys my mother would call "rascals."

"Go back home," one of them yelled at us, and the others laughed.

A couple of them tried to block our path.

My reaction was instantaneous: say nothing, avoid eye contact, and hope at least to avoid an open confrontation. I grabbed my mother's arm, looked down, and tried to maneuver quickly past the boys.

For much of my life, my mother seemed the model of compliance and obedience. Quiet, generous, and deferential, she did all the cleaning, cooking, shopping, and laundry for the family, constantly worrying about all of her children.

I worried about her, too. Even as a child, I felt protective of her and how she was navigating a world that was new to her, raising four children in a country that was not her own. A place where she did not, at first, speak the language or know anyone. A small woman in a sari, her hair always neatly pulled back into a single plait, she was not exactly the standard-bearer for what we typically think of as "defiance."

And yet, she could—and still can—surprise me.

In that moment in the alley, she shrugged me off and, with a firmness that shocked me, pulled the shopping cart vertically upright. Her face was set in a determined scowl, her hand on her hip in a half power pose. She is a petite woman, four foot ten at most, but she suddenly looked taller.

"What do you mean?" she asked in a clear, uncharacteristically strong voice.

"Come on, Ma," I whispered to her.

"Oh yes, you think you're clever?" she said, staring the boys down.

The boys seemed cowed now, their eyes darting to each other. None of them answered.

"You think you're so strong," my mum continued. "Big tough boys, right?"

Then one of the boys muttered, "Let's go."

And they dispersed.

Mum carried on walking through the snicket, head held high, her green sari trailing behind her in the breeze. I stood frozen for a few seconds, then ran to keep up beside her.

My mother made this trek twice a week by herself, and although she occasionally came into the house muttering angrily to herself, I always assumed it was because she was frustrated with our unwieldy shopping cart.

But this had clearly happened before. She had encountered these boys—or people like these boys—for years.

And that day, something changed. She was tired of giving in. She had had enough, and she spoke up.

I had never dreamed that my mother would tell off a group of street toughs. It was so far beyond her normal behavior that until it happened, I didn't even think it was within the realm of possibility. She had always seemed to me a compliant person—someone who took pains to defer to others, who was unaggressive and unassertive, who always kept quiet and served the needs of those around her. I had slotted her neatly into a binary definition of obedient and defiant people.

But I was wrong. I learned that day that it was not so simple, that defiance sometimes comes from the people you least expect. And that though it often looks like an instantaneous reaction, its progression—from feeling to intention to action—can take a long time.

My mother's defiance was a surprise to me. But I do not think it surprised her—she was ready for it.

Defiance Is a Practice, Not a Personality

When I started writing this book, my editor asked me to name someone I'd always thought of as a "defiance role model"—someone from my life who embodies the spirit of what it means to defy. No one came to mind immediately, and after a few days I realized I didn't have a person in mind, but rather a set of defiant actions. The image of my mother berating those boys in the alley in Yorkshire is etched into my memory with striking clarity. Equally indelible is a story about my father. Although he was the authority figure within our family during my childhood, in my eyes he was also the epitome of compliance and quiet reservation outside the home.

When I started high school at St. Joseph's across town, there was no direct route from my house. I had to take one bus, walk a few minutes to another bus stop, and catch another one. A round trip of four daily buses was a significant expense.

At my dad's instruction, I applied to the town council for a subsidized bus pass. It felt like a formality, as all the children I knew that had to travel to school had been granted one.

But one day, a few weeks later, a letter for me came in the mail. I opened it in the crowded living room and amongst the chatter of my siblings read the verdict out loud:

"Your request for a bus pass has been denied."

My dad looked startled and asked me to repeat what I had said.

"Your request for a bus pass has been denied."

My dad asked to see the letter and then stared at it for a long time. He didn't say a word.

But later that night, after I had gone to bed, he started building a case for why I deserved that pass. He worked on it in the evenings, after work, for nearly a month. And at the council's next meeting, he presented it.

I wish I could have been there to see him that day. My father had never attended a council meeting, and I couldn't imagine him speaking at one. But I heard the story later: that when it was his turn to speak, my father stood up, his notes in his hand, passionately emphasizing each of his key points. After a few minutes, he put his notes neatly into the pocket of his suit jacket, deciding he didn't need them anymore. The council members' denial of my bus pass, he argued, was against the ideals of British society, the responsibilities of a state to its citizens, and "perhaps even against the laws of the land"—this last point a rhetorical flourish that made one of the town council members smile.

The council was not used to such passion and defense of civic ideals. They politely thanked my father for his time and informed him that he would hear their decision in due course.

When I received my bus pass a few weeks later, I knew what my dad had done for me. The image of my immigrant father, confronting the large panel of white faces of the town council, resonated within me like a struck bell. I felt proud and liberated—a bureaucratic structure that had seemed so immovable and set against us could be challenged, successfully. This time we didn't have to accept the low hand we'd been dealt.

Both my mother's and father's positive defiant actions I witnessed as a child have stayed with me. They stand out, like my friend Clara waltzing into the school assembly without her red sweater. These experiences made me realize that defiance is not an inherent character trait, nor a consistent identity. You can be defiant one day and compliant the next. Defiance is an *action,* a variable behavior that depends on the situation. Just as there are no inherently "good" or "bad" people—only moral or immoral actions—there is no such thing as a completely defiant person, or a completely compliant one. The movement from compliant to defiant is always ongoing. But these moments of defiance, observed in others or in ourselves, have the power to shift our core sense of self—what we understand, what we value, and what we perceive as possible. They shape our ideas about the type of society we'd like to live in.

This might not have been what my editor wanted to hear when she asked me to name a defiant role model. But I find this idea enormously freeing. It means that defiance is possible for anybody, even the people who consider themselves compliant.

We may be wired for compliance, but we are not trapped by it.

Compliance might be our default, but like a muscle, we can build up our ability to defy. When we understand the pressures that keep us compliant, and what our True Yes really is, we can begin to do something different. It takes understanding, self-awareness, effort, and practice, but we can disentangle the idea that *compliance = good, defiance = bad.*

We can begin to recognize the situations in which compliance is bad and defiance is actually *good.*

We can change our wiring.

Part Two

A TRUE NO

6

Find Your True No

On December 1, 1955, a department store seamstress boarded a city bus in Montgomery, Alabama. She found a seat in the middle section of the bus, next to a Black man sitting in the window seat and across the aisle from two other Black women.

At the third stop, the bus became so crowded that there were no more seats left in the front, and a white man was forced to stand. Seeing this, the driver called out to the four Black passengers seated in the first row of the middle section to give up their seats. As the bus was full, all four passengers would have to stand so that one white man could sit down.

At first, no one moved. Then, after further prompting from the driver, three of the passengers got up and did what the man told them to do.

But the seamstress stayed in her row, sliding over to the window seat.

The driver walked down the aisle to her. "Are you going to stand up?" he asked.

"No," she answered.

"Well, I'm going to have you arrested," the driver said.

"You may do that," she replied.

Years later, in her autobiography, Rosa Parks would write:

> People always say that I didn't give up my seat because I was tired,
> but that isn't true. I was not tired physically, or no more tired than
> I usually was at the end of a working day. I was not old, although
> some people have an image of me being old then. I was forty-two.
> The only tired I was, was tired of giving in.

There's a reason why schoolchildren across the United States are taught about Rosa Parks. Her actions are textbook defiance: principled, high-stakes, and an effective catalyst for lasting change. Her refusal to give up her seat that day in Alabama helped spark the Montgomery bus boycott, one of the defining mass actions of the Civil Rights Movement. Yet I'm often surprised by *how* people talk about this icon of defiance.

I was raised in the U.K., far from Alabama, and I don't recall being taught about Rosa Parks at school; I read her book as an adult. But I have always been taken aback by the misunderstanding that she did not want to get up from her seat because she was tired. A *Los Angeles Times* article about Parks, written in 1965 for the ten-year anniversary of her refusal to move, emphasizes how she was "tired all over" after her long day of work. Even her obituary in *The New York Times* in 2005 referred to Parks as the "accidental matriarch of the civil rights movement."

Thinking of Rosa Parks in this way reduces her action to circumstance, as though everything that happened that day was a matter of chance. It minimizes Parks's role in history, recasting a deliberate principled stand as merely a matter of sore feet.

It also dilutes what I believe is most interesting about Rosa Parks's act of defiance: everything that led up to it.

Rosa Parks did not become defiant overnight. As Jeanne Theoharis's book *The Rebellious Life of Mrs. Rosa Parks* reveals, her life was shaped by what she often referred to as "a life history of being rebellious." Rosa's mother, Leona McCauley, a schoolteacher, embodied a spirit of self-determination. Once, in an incident that Rosa would remember the rest of her life, her mother politely refused to move when a bus driver indicated she could not sit next to a white man.

"I'll throw you off the bus," he said.

"You won't do that," Rosa's mother quietly replied.

The driver returned to the front of the bus, and that was the end of it.

Although shy, young Rosa was no pushover—as a child, her strong will sometimes got her into trouble. Once, walking home from school, a white boy on roller skates pushed Rosa into the street. She dusted herself off and immediately pushed right back, sending the boy sprawling into the dirt. The boy's mother threatened to call the police, but backed down when Rosa explained that she was simply defending herself from an unprovoked attack. Rosa's grandmother told her she had to change her approach to white people, or "you'll be lynched before you're twenty."

So Rosa learned how to tread what she called "the tightrope of Jim Crow"—laws that mandated segregation by race. She complied to survive as her grandmother had taught her, while preserving her sense of herself as an autonomous, independent person, as her mother had taught her.

At nineteen, she married Raymond Parks, a self-educated barber and activist ten years her senior, who was fighting to free the Scottsboro Boys, nine Black teenagers who were falsely accused of rape and sentenced to prison and death by all-white juries in Alabama. Raymond and his fellow activists took food to the nine young men in prison, wrote leaflets, raised money, and strategized new ways to publicize the boys' plight. Rosa soon joined their meetings, which were so secret that no one used their real names, and so dangerous that the table of Rosa and Raymond's house was often piled with firearms for protection.

Two of Raymond's fellow activists were killed for their activities, but that did not deter the rest. They kept fighting. In 1943, Rosa joined Raymond at a local NAACP (National Association for the Advancement of Colored People) meeting. As the only woman present that day, she was asked to take notes and then was elected for branch secretary as she was "too timid to say no." She held that leadership position for the next fourteen years, crisscrossing the state, helping Black people register to vote and pressuring the police to investigate crimes against Black women, which they so frequently ignored. She met with unjustly incarcerated people, attended seminars on civil rights strategies, and interacted with fellow activists, including a young Martin Luther King, Jr.

It is misleading, then, to think of Rosa Parks as simply a tired seamstress. She was in fact a key figure in Montgomery activist circles, and well known for her strength of character and quiet resolve. Her path to bus number 2857 in Montgomery was a long and intentional one.

Her path also had a clear, if underappreciated, precedent. Through her work, Parks mentored Claudette Colvin, a fifteen-year-old-girl. On March 2, 1955, almost nine months to the day before Parks's own stand, Colvin declined to move for a white woman on a Montgomery bus. She was forcibly removed from the bus and arrested for disturbing the peace, violating the segregation code, and assaulting police officers.

Parks herself had long chafed at the city's bus system, preferring to walk if possible to avoid the abuse and slander directed her way by white passengers and drivers. She often declined to board through the back door, as was required of Black passengers after paying their fare at the front. On one memorable occasion in 1943, a driver named James Blake told her to exit the bus and reenter from the rear. When she did not move, Blake tried to physically push her off the bus.

"I will get off. . . . You better not hit me," she told him.

She exited the bus—but before she could step back up into the rear entrance, Blake drove away. Standing in the rain, soaked to the bone, Parks vowed to never ride one of his buses again.

But on December 1, 1955, she found herself, unknowingly, once again on James Blake's bus. And this time, when he told her to move, she didn't budge.

Why?

It was not Rosa Parks's first experience of racial discrimination and hatred. A lifetime as a Black woman in Alabama had inured her to racism, prejudice, and injustice. But just four days earlier, she had attended a meeting responding to the murder of Emmett Till, a fourteen-year-old Black child who was lynched for allegedly whistling at a white woman.

The widely circulated image of Till's mutilated body, presented in an open casket ceremony by his mother, crystallized for her the experience of racial discrimination, pain, and persecution that she and so many other Black people faced. After years of effort, she had grown increasingly despondent about the effects of her activism. The tightrope was fraying beneath her, and she could not bring herself to toe the line for one more day.

"I thought of Emmett Till," Parks would say later, of that day on the bus, "and I just couldn't go back."

Or as the poet Nikki Giovanni wrote, in her poem "Rosa Parks":

> Mrs. Rosa Parks . . . could not stand that death. And in not being able to stand it. She sat back down.

This is what so moves me about Rosa Parks's defiance: Her refusal to give up her seat was emotional, but it was not dictated by emotion. It happened quickly, on a day like any other—but to view it as spontaneous is a mistake. Her dissent wasn't a reaction; it was an *action*. It wasn't an impulsive response, but a decision decades in the making, the end result of a lifelong process. Whether she knew it or not, Rosa Parks had been practicing for that day all her life.

Consent and Defiance: Two Sides of the Same Coin

Rosa Parks's refusal to move in 1955 was clearly defiant: a considered, direct, uncompromising action derived from her core values. It was her True No.

Defiance and consent might appear, at first glance, to be positioned at opposite ends of a spectrum, but in fact they are more similar than we might think. Like consent, defiance requires the same five elements: capacity, knowledge, understanding, freedom, and explicit authorization. And like a True Yes, a True No is a *thoroughly considered action, based on our most authentic values.*

The figure on the next page (figure 2) illustrates the similarities of the journey from mere compliance to either a True Yes or a True No. Compliance is where most of us start—the most *closed* position on the figure where we are most unenlightened. As we move through the elements, we *open* our minds to the possibilities of true consent or defiance. When we consider our values and whether we have the capacity (the ability to make a sound decision), knowledge (the information necessary to make it), understanding (comprehension of that information), and freedom to decide, then we can authorize a True Yes or a True No.

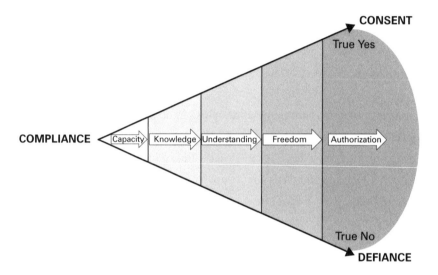

Figure 2: From Compliance to Consent and Defiance

Whether you say the word "yes" to give your informed consent or "no" to give your informed refusal is largely situational, depending on what is being asked of you or what is expected from you. But in a way, it does not matter whether you say "yes" or "no"—what matters is that your decision helps you live in alignment with your values.

Rosa Parks knew she had the *capacity* to make a decision that day on the bus. She had a lifetime of *knowledge,* not only of the rules and laws governing her behavior on the bus, but of the larger social context. She *understood* what would happen to her if she refused to move—the risks, that she likely would face arrest and imprisonment, and the benefits, that her refusal could be a symbolically powerful direct action against Jim Crow. And she realized she had the *freedom* to make this choice, even though it would come with consequences. Prepared and willing to accept the costs, Rosa Parks gave her True No.

Stages of Defiance

The more I've thought about defiance—from my experiences as a child growing up in Yorkshire, to my time as a psychology student poring over the behavior of the Milgram participants, to my years as a physician and organizational psychologist conducting hundreds of interviews and observations—the more I've come to realize that reaching a True No is a process, not a snap judgment. Milgram himself describes an escalating sequence in some of his defiant subjects. A progression through a series of stages, built on past experiences, deliberation, and self-connection (see figure 3).

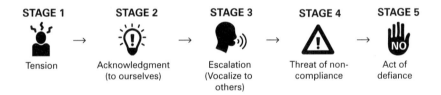

STAGE 1	STAGE 2	STAGE 3	STAGE 4	STAGE 5
Tension	Acknowledgment (to ourselves)	Escalation (Vocalize to others)	Threat of non-compliance	Act of defiance

Figure 3: The Five Stages of Defiance

We are most familiar with the first stage—the uncomfortable feeling of *tension* that arises in the light of unwanted influence. Tension emerges when our core values collide with the needs or expectations of others—it is the signal that it may be time to resist.

Stage two is the *acknowledgment* and recognition of that tension within ourselves. Acknowledging our tension reminds us of our agency—if you had already handed over all your power to someone else, you wouldn't feel any tension. When Thomas Simms, the McDonald's handyman, was ordered by Officer Scott to ask Louise Ogborn to remove her apron—the only item of clothing she had to cover herself—he immediately felt and acknowledged his discomfort. Recognizing our discomfort is critical to being able to articulate it to ourselves and others.

Stage three is the *escalation* of our internal discomfort by expressing it externally. This is a significant stage. In the Milgram experiments, participants who refused to administer shocks often registered their discomfort to the experimenter early in the process, as soon as they recognized and acknowledged the tension in themselves. Those who remained silent were more likely to keep administering shocks, perhaps thinking their time to resist had passed. Vocalizing our discomfort to others reduces the pressure to rationalize our complicit actions at a later point and increases the likelihood that we will move on to the next stages of defiance.

Such vocalization need not be direct or confrontational. We may at first subtly question, as the Milgram's participants did: *Are these shocks safe? Is the learner okay?* Or, as I did about my unnecessary CT scan: *It's only a small amount of radiation, isn't it?*

At this stage, we are still maintaining a deferential position. If our concerns are dismissed, we need to repeat them, to let others know that we are not comfortable with what is expected of us.

Roger Boisjoly, the Morton Thiokol engineer who warned of disaster prior to the *Challenger* mission, expressed his concerns about O-ring function in cold weather months before the launch. In one memo, he warned, "The result could be a catastrophe of the highest order, loss of human life." When his warnings were ignored, he intensified his protest. On a conference call with four other Thiokol engineers and NASA the night before the launch, he yelled, "How the hell can you ignore this?"

When we outline our concerns to others and continue to do so when dismissed, we are laying the groundwork for stage four: *a threat to stop complying.* Such threats often precede the final stage: an *act of defiance,* a True No.

Feeling tension, acknowledging it to ourselves, escalating concerns to others, and threatening not to comply brings our will to defy out of the realm of inchoate feeling and into the world of language and action. Progressing through these stages is often difficult. But in my own experience and those of others I've interviewed, and as Milgram also noticed in his subjects, the final act of defiance is often an affirmative one. It's a True No to the situation and a True Yes to our values. And the striking thing is that once we act in alignment with our true values, much of our tension, anxiety, doubt, and fear evaporate. Defiance can be a relief.

These stages don't always occur in order. They may show up simultaneously, or you might skip a stage—not realizing how tense you are or neglecting to express your concern to others—only to return to it later. You might toggle between two stages, back and forth—recognizing and acknowledging your tension, then escalating, but stopping short of a threat of noncompliance, then returning to sitting with your tension, and either ending there or escalating once more. You might not experience each of these stages—you may jump from stage two to stage five, as Thomas Simms did, without ever experiencing stages three and four.

But the stages are a useful model for understanding how defiance can function as a progression. Many of these stages don't necessarily look like our iconic images of defiance: that's the point.

What we think of as "defiance"—Parks refusing to move on the bus— is in fact the culmination of a process. And just as important as the final act of defiance are the stages leading up to it.

We can see in Parks not only the world-changing potential of her defiant act, but the slower, more patient, and transformative effect of a lifelong training in defiance: from her childhood in rural Alabama, to her work fighting for the Scottsboro Boys, to her efforts registering Black people to vote in one of the most highly segregated cities in the country. We can see it all unfolding through the course of her life: her tension, her acknowledgment, her escalation, her threat of noncompliance, her *defiance*.

Most of us have not had a "Rosa Parks" moment. We get stuck well before we get to stage five, the defiant action we wish to take. Sidetracked by our wiring for compliance, discouraged by our tension and discomfort, or stymied by circumstances, we are often unwitting compliers in our own lives. But that doesn't mean we are not capable of defiance. It just means that we haven't yet learned how to harness our tension in service of our True No. We might have felt the urge to defy, but we just haven't—as yet—learned to complete the process.

The time frame for defiance can be long. But we can learn how to move from one stage to the next, and—crucially—we can practice, so we know how to speed up the process of defiance in the moment. And this, to me, is one of the most encouraging things about true defiance. Though it is often extraordinary, it is not the sole province of extraordinary people. I have always been inspired by Rosa Parks not because she made a spontaneous decision, but because she had prepared for that crucial moment for most of her life. She was ready when that decision came—not because she was superhuman, but because she was eminently *human*.

Rosa Parks is a towering figure in American history. But she was also a person in the world, just like you and me—an ordinary person who made an extraordinary decision.

7

The False Defiance Trap

On January 6, 2021, President Donald Trump told the assembled crowd at the "Save America March" that the election should be overturned and his people needed to show strength and fight for their country. Thousands of supporters then followed the president's wishes and marched down Pennsylvania Avenue and the National Mall to the Capitol. Many of them then bypassed security, attacked police officers, and occupied the legislature's central building for over four hours. Hundreds were arrested, hundreds were injured, and five people died.

President Trump was later impeached by the House of Representatives for inciting an insurrection. And although a number of people in the crowd that day were later convicted of planning their attack in conjunction with far-right hate groups and militias, the majority of those charged with crimes in relation to the day's events were unaffiliated with any such organization. Some of these people claimed to simply be following the crowd—they had just got caught up in the moment, they said.

One such person was Clayton Ray Mullins, a fifty-two-year-old white man and devout Christian who ran a salvage yard and tended to a tiny Baptist church in rural Kentucky. He had voted for President Trump, but he was not known in his hometown as a particularly political person, and he himself believed that the election had been a fair one. "No one man has the power" to overturn an election, he said later, referring to Trump's repeated claims that he had in fact won. "You're not supposed to put one man up on a pedestal and think he's going to bring peace to the world."

And yet, when Donald Trump told him to march, he did. When the raucous crowd burst through the barricades at the U.S. Capitol, he climbed atop a balustrade, holding an American flag. And when the Capitol Police attempted to push the crowd out of the government building, Mullins pushed back.

Clayton Ray Mullins is now one of many people sentenced to prison for his actions on the day of the Capitol riot—video footage from that day captured him dragging one police officer down the stairs and, as part of a crowd, pushing others. In September 2023, he pled guilty to the felony offense of assaulting, resisting, or impeding officers, and in January 2024, he was sentenced to thirty months in federal prison.

What makes a man like Mullins, by all accounts mild-mannered and abstemious, act the way he did? How does a peaceful and kind man, as described by friends and family, end up in a violent crowd?

A Defiance Mirage

Whatever made Mullins act the way he did that day—loyalty, peer pressure, herd behavior, or some combination of these—it was unlikely to be true defiance.

It's important to note that defiance is not the province of any one value system, political party, or ideology. Just as some of the January 6 rioters were pushed along by the crowd rather than their own values, there were undoubtedly many people pushed along in other events like the Women's March in January 2017, wearing knitted pink hats more out of conformity than true defiance. False defiance is something any of us from any ideological perspective can slip into; that's what makes it so dangerous.

And while some people at the Capitol insurrection may have been defying in alignment with their values, Clayton Ray Mullins, by all accounts, was not one of these people. He didn't even particularly want to attend the rally—he went because his wife and his sister did. That day at the Capitol, his actions did not match his professed Christian values or his reputation for fairness and peace. In the moment, his independent self, the part that believed the election was fair and violence was unnecessary, gave way to his interdependent self, which wanted to match its behavior to the crowd. He was drawn into the crowd's behavior even though, as he confesses, he did not fully succumb to their ideology. For a few hours, he might have looked defiant, on security footage and television screens, where he quickly earned the nickname "Slickback," for his signature coif.

But when he was out of the crowd, his independent self resurfaced. Walking back to his car, he wept. On the roughly eight-hundred-mile drive home with his wife and sister, as the view from the car window changed from the urban bustle of Washington, D.C., to the Blue Ridge Mountains of Virginia and finally the rolling green hills of Kentucky, he did not say a word. He said he felt like a man waking up from a dream.

Crowds and group dynamics don't always enforce false defiance— sometimes they can do the opposite. Milgram saw this firsthand: In one of the variations of his shock experiments, the teacher, the participant administering the shocks to the learner, was joined by two other "teachers" who, unknown to the main teacher, were actors. Each of these actor teachers had been instructed to refuse to proceed with the experiment; the first at 150 volts—when the learner gave his first vehement protest to be released from the experiment—and the second at 210 volts.

Under these conditions, the behavior of Milgram's study participants changed drastically. They became much less compliant. When the teacher was exposed to the defiant actions of their peers, their own defiance increased to a dramatic 90 percent.

Support from others can impressively undermine pressure from authority. People feel much more comfortable defying when at least one other person is visibly doing so. In some ways, this is unsurprising: marches, sit-ins, and protests often inspire a group of people with a similar vision, objection, or common purpose. There is strength in numbers; often our True No is more easily accessed when it is echoed and amplified by others sharing the same values.

But the presence of other people can also make it more difficult to know the precise motivations for our actions. If you are the *only one* wearing a mask at an indoor gathering during the COVID pandemic, you are likely to be expressing your true values. But if everyone on your block is already displaying the same "Women's Rights Are Human Rights" or "Proud to Back the Blue" lawn sign as you are, then is your sign an act of conformity, compliance, consent, or defiance? Similarly, when you are marching or holding a picket sign alongside hundreds or thousands of others, how can you be sure that your True No is in fact what you are expressing?

Recognizing False Defiance

Behavior like that exhibited by Clayton Mullins might *appear* defiant, but if it is disconnected from a person's true values and higher principles, it is a false defiance, a flimsy imitation of the real thing, and it is not limited to the political arena.

You probably see this every day—especially if you have children.

I have certainly had this experience with my teenage son. Dinner is over, the plates are in the dishwasher, the computer and his phone are within reach.

"Do your homework first before playing," I say.

"I was just about to," he responds.

A minute passes, then two, then fifteen. The eyes stay glued to the screen; the fingers dance over the buttons.

"Are you going to do your homework?"

"I was just about to," he repeats. "But now that you want me to, I'm not going to do it."

I know that his actions are not, in fact, defiant in line with my definition. A child's knee-jerk reaction to oppose a parent's will is rarely the result of their own principles or values. It is just a belligerent reflection of the parent's own values, given in opposition. It is not a proactive decision, but a reactive position. My son's action shows that he is clearly listening intently to my wishes, not ignoring them. His "no" is wholly dependent on my preferences.

When I was a teenager in Yorkshire smoking a cigarette with the door closed, I wasn't exhibiting true defiance. Lighting up wasn't an expression of my nature, or an enactment of my principles. I wasn't smoking that cigarette for any deeply considered reason. I was just doing it because my friend Clara did. I was simply copying someone else. That's why the habit didn't take, and why my father so easily dissuaded me from continuing it.

Many of us engage in this kind of false defiance as adolescents as a way to understand who we really are. It is difficult to arrive at a deep knowledge of our true values, and often, especially when we are young, we "try on" various ways of moving through the world, embodying others' behaviors and values to see how they feel. That isn't to say that adolescents are wrong to question authority or that their defiance is always false—sometimes teenagers can quite perceptively point out painful realities to us. And like adolescents, sometimes adults do not know, or they forget, what their values are.

Algorithms on social media encourage false defiance because extreme views, polarization, reflexive opposition, and "takedowns" get the most views. On the internet, the performance of defiance is often more tantalizing than substantive, transformative defiance. Going viral by making others uncomfortable can feel good, especially when the cause is an admirable one. And so can being seen as "right" rather than actually doing the right thing.

True defiance is about living a life aligned with your values, not scoring points by virtue of sheer opposition. In contrast, false defiance reflects or reacts to the values of others; it does not embody your own principles. The problem, as Clayton Ray Mullins found when he faced a range of felony and misdemeanor charges, is that just as compliance often looks like consent, *false* defiance often looks real to other people. Whatever his true intention that day, what most people saw was simple: Mullins on the Capitol steps, holding the boot of a police officer and pulling him into an unruly mass. Observers are likely to think of him through the lens of their political convictions, either as a member of a treasonous mob or as a patriotic hero, protecting democracy—when in fact he may have been simply complying with what he was told to do and conforming with the pull of the crowd.

True defiance can be so difficult, its costs so high, that many people often defer it to another day. One can imagine that someone like Mullins might look back at his actions at the Capitol with a kind of disbelief, recognizing his behavior as unrepresentative of his core values. How dispiriting, then, to assume all the risks and costs of defiance without any of the values underlying it—to expose yourself to the negative consequences that keep people compliant without experiencing any of the world-changing possibilities of true defiance.

"We never should have come here," Mullins said to himself as he trudged away from the Capitol building after the riot.

The Two Sides of Moral Convictions

Our ethical values often empower us to act defiantly. They push us to do what we think is right, regardless of the opposition. In one variation of Milgram's shock experiments, a divinity school professor, an expert on the Old Testament, relied on his core values to defy the experimenter. In this particular study, the setup was the "proximity version," in which the learner was placed in the same room as the teacher, just a few feet away. This closeness brought the learner into focus for the teachers, and increased defiance. When teachers could see the learner in addition to hearing his complaints, defiance rose from 35 percent to 60 percent.

In further setups of the proximity version, teachers were required to physically connect with the learner by forcibly placing the learner's hand on the shock plate after the learner refused to do so. Defiance increased even further in these cases, to 70 percent. The more teachers could hear, see, and feel the learner, the more "real" and "human" the learner became. It is difficult to dehumanize or ignore the harm someone is inflicting on another person when one is face-to-face with them.

What's interesting about the divinity school professor's defiance is not only what he did but also what he said. The professor first positioned his chair so that he was directly facing the learner while administering the electric shocks. When the learner made a mistake with the assignment, the professor's corrections were like those of an irritated tutor to a somewhat subpar student.

Still, at the 150-volt mark, when the professor saw the learner writhing in pain in front of his eyes and shouting out in agony, "Experimenter, get me out of here! I won't be in the experiment anymore! I refuse to go on!" something changed for the professor. He also refused to go on; he declined to administer any further shocks.

When the experimenter told the professor that it was essential to the experiment that they continue, the professor replied unequivocally: "I understand that statement, but I don't understand why the experiment is placed above this person's life. . . . If he doesn't want to continue, I'm taking orders from him."

The professor's words revealed his value for *humanity*, and for the sanctity of human life. He did not wish to inflict harm on another person.

Further instructions from the experimenter—"You have no other choice, sir, you must go on"—did not sway the professor.

"If this were Russia maybe, but not in America," he responded. Although the professor's voice trembled and quivered, he treated the experimenter, Milgram wrote, "as a dull technician who [did] not see the full implications of what he [was] doing."

"If one had as one's ultimate authority God," the professor said in an interview after the study was completed, "then it trivializes human authority."

The professor's words suggest that he wasn't so much defying orders from the experimenter as he was following them from someone else. At first, those orders were from the learner: *If he doesn't want to continue, I'm taking orders from him.* Then, the professor invoked God as the relevant authority figure whom he must obey. Perhaps defiance had a negative connotation for the professor, and suggesting his actions were compliant to another authority gave him cover.

But the professor's words also reveal something deeper. As well as disclosing his value for humanity during the experiment, the professor's statements afterward suggest he might follow whatever he believes is God's wish.

God's wish might, for example, encourage the professor to believe that *homosexuality is wrong,* or that *abortion is wrong.* When we have strong feelings about something, whether they come from a religious authority, a political affiliation, or membership in a group, we can become convinced that our beliefs are objectively true. Psychologists refer to these types of beliefs as "moral convictions": unshakable beliefs that something is right or wrong, moral or immoral.

Moral convictions can be so deeply felt that they enable us to oppose authority, ignore pressures to comply, and even break the law. When we feel that something is morally wrong, we are less susceptible to external influence, and more willing to take a stand for our beliefs.

Such convictions can give us strength and the motivation to defy. But unexamined, such convictions can be dangerous. If the only belief that matters to us is our own subjective sense of what is right and wrong, we lose tolerance for different opinions, and our convictions can drive us to commit acts that instigate divisiveness and destruction.

On May 31, 2009, George Tiller was where he always was on a Sunday morning: in the foyer of the Reformation Lutheran Church in his hometown of Wichita, Kansas, handing out programs to the congregation. A sixty-seven-year-old with soft silvering hair, wire-rimmed glasses, and a gracious Midwestern demeanor, Tiller was a father of four, a grandfather of ten, and a man of deep faith.

Also at church that morning was a relatively new attendee, fifty-one-year-old Scott Roeder, who called himself a born-again Christian. Entering through the door from the sanctuary into the church foyer, Roeder walked up to the refreshment table where Tiller was standing, pressed a gun directly against the side of Tiller's forehead, and fired.

Some said later that the sound didn't register as a gunshot—just a loud pop. Gary Hoepner, a fellow usher, initially didn't think it was real—until Tiller fell to the floor. And then there was chaos.

Several witnesses chased Roeder into the parking lot, stopping only when he threatened to shoot them, too. As he climbed into his car, one called after him, "How could you do that?"

"He was a murderer!" Roeder shouted back and drove off.

George Tiller was the son of a physician, and as a young boy, he had tagged along with his father on house calls, carrying his medical bag. Tiller went on to medical school himself and later took over his father's practice. He was well known, liked, and had a lot of support in his community—but he also had his detractors. Tiller ran a clinic that served patients seeking legal "late-term" abortions. He performed his first abortion in 1973, when *Roe v. Wade* hit the books. By the mid-1980s, he was performing more complex terminations for women who had learned after twenty-one weeks, the accepted threshold of viability, that their fetuses had severe abnormalities or were brain-dead. One of only a handful of doctors in the entire country willing to perform these late-term abortions, he became a mecca for desperate patients. But his practice also made him a target.

At no point did Scott Roeder, during his trial, conviction, or sentencing, deny pulling that trigger. In court, when his lawyer asked, "Did you go and shoot Dr. Tiller?" Roeder replied, "Yes."

Yet he also pleaded not guilty to first-degree murder. He insisted that he was acting to defend unborn children. He described how through his religious faith, he'd come to understand how wrong abortion was, that "It is not man's job to take life—it's our Heavenly Father's."

There is both power and danger in moral convictions. The same independence from authority that our convictions can provide can also lead to an attitude wherein the ends justify the means, where getting the "right" outcome, according to one's own beliefs, is more important than how it is achieved. A strong moral conviction that something is right or wrong (e.g., *abortion is wrong*) is different from a core value. Values such as *integrity, compassion,* or *humanity* cannot be classified as facts. Moral convictions, on the other hand, are experienced in the same way as a fact—an objective, irrefutable truth. The problem is, strongly held moral convictions usually have two possibilities about what is right or wrong. They are subjective.

Although a focus on how things "ought to be" may protect us from pressure from authority and even the rule of law, moral convictions can also make us tolerant of nearly any means, including lying, violence, and murder, to achieve what we believe is a morally preferred end. When we cross that line, one person's right is another's wrong.

Moral convictions can sometimes help us achieve true meaningful positive defiance, as they inoculate us from pressures to obey authority. They may even align with our core values, as with the professor who revealed his value for humanity and his conviction that harming another person is morally wrong. But true defiance doesn't proceed from any sectarian religious beliefs, ardent nationalism, or political affiliations; it is not the result of membership in any opposition movement or group. Danger arises when we become so obedient to a belief that something is right or wrong that we abdicate our responsibility to evaluate our moral stances against our actions to achieve what we believe is a desirable end.

False defiance is fool's gold: It might sparkle like the real thing, but it lacks the value of a True No. Real defiance is a thoroughly considered decision that aligns our actions with our core values. This action may sometimes align with our moral convictions, but it does not stem from a cause or movement. It may sometimes happen in concordance with others, but it does not flow from blind obedience or loyalty. It is an individual choice; not a crowd-sourced reaction; a considered decision, not an instantaneous or performative response.

We enact a True No not because we aren't thinking—but because we *are*.

8

Who Gets to Defy?

The first time Kenneth got pulled over by the police was just one week after he got his driver's license. At sixteen years old, he was a young, fresh-faced high school student with ambitions to attend Berkeley. It was a perfect California day: a Friday in late spring, warm and sunny with clear blue skies. After his last class, Kenneth walked quickly to the parking lot and hopped into his silver Chevy Malibu, delighted that the weekend was ahead.

Kenneth turned up the music—Kendrick Lamar, *good kid, m.A.A.d city*—and began the short drive home. Just two blocks from the school, as he pondered his evening plans, he heard the abrupt *whoop* of a siren, then saw the bright red and blue flashing lights of a police car in his rearview mirror.

Kenneth had no idea what he could have done to warrant a traffic stop, and his heart immediately started racing. But he turned on his hazard lights, slowed down, and found a spot on the side of the road to park. Then, although he was terrified, he began to follow the script his father, a twenty-year veteran of the police force, had first given him just a few months ago after he got his learner's permit: He rolled down the windows, put his hands on the wheel, and reminded himself to keep his voice calm and moderate, to say "Yes, sir; no, sir" or "Yes, ma'am; no, ma'am" when addressing the officers, and above all, to make no sudden movements.

These were important things to do, Kenneth's father had told him, and not for moral or ethical reasons. They were important for his own safety. Kenneth, like his father, is a Black man, and even a minor traffic stop could turn out to be more dangerous for him than for others. "The Talk" he gave Kenneth was much like the one his own father, also a veteran of the police force, had given him as a young man: to obey the police, to not say or do anything that could appear potentially threatening, to comply with any request, even if he disagreed with it.

Kenneth watched in the rearview mirror as two white officers, both with sunglasses, approached the front car windows. Each of them touched the back of his car. Kenneth knew why—if he fled the scene, investigators could tie the officer's fingerprints to the vehicle.

One of the officers stared into the back seat, as though expecting to find something there. The other rapped on the roof of the car and asked Kenneth if he knew why they had pulled him over.

"I don't know, sir," Kenneth answered.

"California stop," said the cop to his right. "License and registration, please."

A California stop, in which a driver slows significantly but does not fully stop their vehicle at a stop sign, is a minor traffic offense, and Kenneth was sure he had not committed one.

Kenneth felt a tight grip in his stomach—stage one of defiance. He acknowledged it—stage two. Although fear was his overwhelming emotion, he managed to surreptitiously check the area for traffic cameras, to see if he could disprove the charge with video evidence. But he saw none. He knew that he should not have been stopped. There was a very brief moment when Kenneth considered questioning the officers or claiming that they were mistaken.

But Kenneth decided against escalating and vocalizing his discomfort to the police officer, the third stage of defiance. He did not refuse to hand over his license or threaten in any way not to comply. In his mind, he heard the voice of his father, telling him not to question authority in the moment, not to raise his voice, to do whatever it took to get out of the situation as safely and quickly as possible. Though his hands trembled, he reached slowly into his glove box for his license and registration as requested, describing each movement beforehand—*I'm taking my hand off the wheel, I'm going to reach across the passenger seat, I'm opening the glove compartment now.*

The rest of the interaction was short. The two police officers wrote the ticket and wished him a nice day, and Kenneth drove home in the blazing sunshine. When he got home, he was still trembling.

For Neil, a tall Black business school student in my class with a wide, ready smile and a thoughtful manner, there was no single "talk." Neil grew up in Newark, New Jersey, where the knowledge of how to behave around the police was just part of the underlying knowledge of his life. The messages came to him from all around. He heard them at the barbershop, at church, at the corner store, at cookouts—in the atmosphere that he breathed in every day. When I asked him what he knew about compliance, he said there were certain things you just knew you needed to do if you encountered a police officer. "One is to always comply. No matter what the situation is, always comply."

In the presence of police officers, Neil knew he would have to codeswitch, changing the way he spoke, choosing his words carefully, and modulating the pitch of his voice. Like many of his peers, he did this all the time in predominantly white spaces. But with the police, the stakes were especially high. He couldn't use the same words or tone with a cop as he could with his friends, for fear that he would appear threatening, dangerous, or suspicious to a white officer.

Once, on the way home from a high school party, Neil was riding in a car with other students that was pulled over by the police. The police believed the students had marijuana in the car even though nobody was smoking. Everyone was ordered out to the sidewalk and to show their driver's licenses. Neil was the only non-white passenger. Terrified, he put his hands on his head, followed every directive issued to him as he had learned, and consciously changed his vocal tone and inflection to appear nonthreatening. But he was amazed to see the relative freedom exhibited by his classmates. Without hesitation, several reached for their phones and called their parents for help.

"My white friends [had] more free will with what they could say," he said. "They don't have to code-switch. They weren't afraid in the same way. They don't feel the same danger."

Their ability to defy, to ask questions, to object, was a privilege, Neil realized. These other children not only had parents who could conceivably get them out of a ticket, but they also didn't even have to think twice about reaching into their pocket to call those parents, just three feet from a police gun.

No one was arrested that day, but Neil never forgot that afternoon on the New Jersey Turnpike. Now, whenever he hears a siren or sees a police officer, his fear rises to the surface. His mind goes to the worst-case scenario, and he knows that in that moment there is nothing he can do with his feelings.

"I want to say something, but do I ever?" he said to me. "No! No, when I get pulled over, I comply. I comply because I want to go home."

Neil's and Kenneth's stories are just two of millions. They're indicative of the way compliance, consent, and defiance are integrally connected with issues of power, authority, and race. Black people in the United States are significantly more likely to be killed than their white counterparts during encounters with the police.

A hierarchy of defiance is conspicuous in our world, governed by social norms, stereotypes, and societal and cultural expectations regarding acceptable behavior. This hierarchy dictates who gets to defy and who doesn't, who has no choice but to be humble, and who is allowed to demand respect. It allows some to defy with few consequences, while saddling others with outsized and often dangerous penalties.

Defiance is not always a safe option, particularly for those targeted in an unequal power structure: for people of color, for women, for queer people, for anyone who does not fit the dominant paradigm. The Talk reveals one way that defiance and compliance function in the real world, where the risks are disproportionately higher for some people than others, and where social structures, hierarchies, and power affect us all in different ways. It illustrates the complexity of defiance, its limits and its possibilities, and reveals the ways in which it is more readily available to some people than others.

As well as the question of who gets to defy, we must also consider: Who is it easier to defy *against*? People often find it easier to defy against those from the nondominant class, who are, in turn, often *expected* to comply. I have witnessed this firsthand many times in academia, where women and people of color are regularly confronted and challenged—sometimes in the classroom, by students, or even by their colleagues and assistants—in ways that their white male counterparts are not.

We all have to weigh the rewards of defiance with the tangible risks: to our psychological well-being, to our livelihoods, and even to our lives. But the truth is that some of us have always had to risk more than others.

The Entitlement of a Tall White Man

Many years ago, as a postdoc living in Chapel Hill, North Carolina, I was chatting with a PhD student, one of the few other women academics of South Asian ancestry at the university. Out of the corner of my eye, I saw a senior male professor, a white middle-aged man who viewed himself as somewhat of a mentor to us, approaching.

"What is it like," he asked, excitedly rubbing his hands together and smiling, "when two people of an exotic race speak to each other?"

The student and I paused in stunned silence. Seconds before, we had been discussing our research, but now all we could do was stare at each other uncomfortably, then at him. I cannot speak for the student, but I know that for myself, in that moment, it took nearly everything I had not to snap back: *What is it like, I wonder, to be able to say whatever comes out of your mouth, no matter how ignorant?*

But I held my tongue, and so did my "exotic" colleague.

Cue my crocodile smile.

I knew, in that moment, that open defiance and confrontation would not do me any good. I was a postdoctoral fellow looking for my first tenure-track professor job, and he was someone whose opinion of me would certainly play a role in determining my academic fate.

So I did what I could to excuse myself from the situation as quickly as possible.

It was not the first time he had referred to my "exotic" race. Although I could not escape the presence of this professor in my workplace, I tried to keep my distance from him.

The PhD student later revealed to me that she spent her weekends babysitting that professor's children—for free. My concerned expression led her to spontaneously rationalize that at least bouncing a ball around with his offspring gave her some physical exercise.

As students and postdocs, we certainly had less power and influence in the university hierarchy. However, this professor treated us differently than our equally subordinate white male peers. In his eyes, we were seen foremost as "exotic" South Asian women. We were expected to be particularly subservient and exploitable in ways that he did not ask of our white male postdoc colleagues. To me, it was apparent that he thought of the university as his, not ours. We were there merely to serve him.

The professor wanted me to be attentive, industrious, and compliant with his needs, but I was not to be in any way forthright about my own opinions and my own needs. Hardworking but not ambitious—or, as Beyoncé put it in "Flawless," quoting Chimamanda Ngozi Adichie: "You should aim to be successful, but not too successful."

At the time, I was older and more experienced than my peers. I was also a new mother myself. I did not, and could not, comply as much as the younger child-free PhD student. My attempts to keep my distance only seemed to infuriate the professor more, and his aggressiveness and objectifying remarks, centered on my race and gender, grew more frequent.

When another South Asian woman academic visited to give a seminar, he texted me during her talk to ask me whether I felt like I was looking in a mirror. Whenever he got the chance to berate me, he took it.

I started to feel sick each time I saw an email from him in my inbox. If I heard his voice in the hallway, I went the other way. I felt tension, and I acknowledged that to myself. But I didn't escalate. I never verbalized my discomfort with the way he treated me, nor did I blatantly defy any of his requests. But I also never went out of my way to please him the way he expected. My refusal to ingratiate myself to him was my form of quiet defiance.

I often wonder what would have happened if, as a postdoc, I had objected publicly to this professor's behavior. What would have happened if my defiance had not been so quiet? Women often experience a harsh backlash when they stand up for themselves. East and South Asians fare the worst in this regard, according to a study by Joan C. Williams called "Double Jeopardy?"—because such behavior upsets stereotypes of their passivity.

The professor's regular use of the word "exotic" to refer to my appearance also objectified Asian women as demure sexual objects and contributed to stereotypes of Asian women as the exotic "property" of white men, there to serve and attend to their every need. He would not have used this word if I were a man.

I was in a hostile, psychologically unsafe environment, the kind so many women and people of color navigate on a daily basis, from the service industry to the C-suite. I knew that objecting would only make my position in that environment less safe. In some environments, we simply need to get out.

Not everyone who experiences such violations chooses compliance. Some people, especially those who are afforded more protection, belief, and support, are able to defy loudly and publicly.

The actor and comedian Mindy Kaling was once asked by an audience member during a panel at the Sundance Film Festival how she had overcome obstacles and succeeded in two industries—television and comedy writing—where the major decision-makers were men.

"That's because of my parents," she answered. "They raised me with the entitlement of a tall, blond, white man."

The line is funny, but her point is serious. The traits that earn many people plaudits in the workplace and are often associated with defiance—independence, forthrightness, even aggression—are evaluated differently depending on what you look like and who you are. People at the top of the defiance hierarchy are "allowed" to be assertive, confident, and forthright with few repercussions. The "tall, blond, white men" that Kaling was talking about are less likely to face serious negative consequences for being outspoken or defiant. They may even be praised for it. They're labeled "uncompromising," or "a leader with backbone." Someone like Steve Jobs was hailed as a visionary and a genius, not in *spite* of his mistreatment of employees, angry tirades, or unreasonable demands, but sometimes *because* of them. When the rest of us behave this way, we run the risk of being called aggressive or arrogant. We may be termed "difficult," "hard to work with," or "not executive material."

One woman I interviewed, Katya, told me about her first client meeting, soon after she was hired at a large consulting firm in Eastern Europe. When she asked her supervisor what her role with clients would be during the meeting, he answered, "Dancing on tables," as the room roared with laughter.

Katya let this slide, although the remark disgusted her. The risks of speaking up were too great. She had only recently signed her offer letter and she too did not want to establish a reputation as being "difficult."

When Katya told me this, I thought of Natalie Portman's remarks at the *Variety* Power of Women Summit a few years ago:

"Stop the rhetoric that a woman is crazy or difficult," Portman said. "If a man says a woman is crazy or difficult, ask him, 'What bad thing did you do to her?' That's a code word. He's trying to discredit her reputation."

Katya was also told that she was irritating her colleagues by asking them about deadlines and goals for projects. She was "a pain in the ass"—just because she was trying to do her job.

"This of course kept me off balance," she told me. "I soon learnt that behaving assertively triggered dislike and was viewed as not feminine enough."

In response, Katya found herself becoming submissive in conversations, undercutting herself and apologizing, even when she had no reason to. Her toxic work culture trapped her in a narrow range of behavior, outside of which she was not allowed to stray. She shrunk to fit the comfort of her colleagues.

Katya's experience is not unique. In so many ways, the risks of defiance have always been greater for those who do not hold power. Moving through life subjected to these discrepancies in risk is like walking on a balance beam. For some, the beam of acceptable behaviors is wide, and they have plenty of room to move from one side of it to the other. They can sometimes comply, other times defy, and no one thinks anything of it.

For others, the beam is as narrow as a tightrope. Lean just a little too far in one direction, and you'll fall right off.

Conscious Compliance

When defiance presents an immediate danger or a significant risk to our well-being, as it did for Kenneth or Neil, compliance can be the most viable survival strategy. Similarly, when we believe that our True No would not be effective or we decide that this is not the battle we want to fight right now—as Katya and I did—open-eyed, intentional compliance can be a way to save our defiance for a more opportune moment.

Most of us have found ourselves at one time or another in an environment where the risks and costs of defiance—psychological, social, financial, physical—are too great. When the costs of defiance outweigh the possible benefits, compliance becomes the tactically best option. Not because we *want* to say a grudging yes, but because doing so is the best option available to us at the time—playing it safe in the moment can help preserve us for more effective defiance in the future.

I call this *conscious compliance.*

Conscious compliance is not consent. It is not an affirmation of our values or dearly held principles.

But it also isn't the kind of automatic compliance that we're wired for, borne out of our cultural or social conditioning. It is, instead, a considered action we make to comply with a situation because the risks of defiance are too great, or the payoffs too meager. We have the capacity, knowledge, and understanding of the situation to know we would rather defy. But we deliberately *choose* to comply (see figure 4).

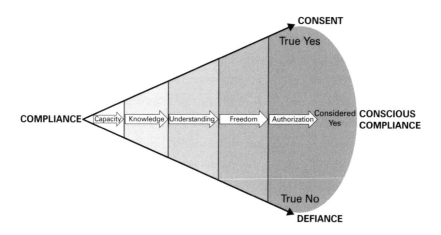

Figure 4: From Compliance to Conscious Compliance

Over the course of my research, I've heard many stories of people who felt defiance was too costly for them, despite their principles. A single mother working her way through nursing school described feeling trapped at her bartending job, when the alcoholic owner demanded that she make him drinks, far past the point of intoxication—she could either overserve him, she said, or she would lose her job. A recent college graduate with a new wife and baby was thrilled to land a position at a top technology company in India—only to find himself stuck between a senior colleague's corruption and his need to keep his job. A young film production assistant spoke of landing her first big feature film, going out to dinner with the male director, and over champagne and caviar, immediately fielding a number of sexually suggestive remarks and jokes she didn't feel she had the power to rebuff.

In each case, the person wanted to defy. The bartender wanted to cut her boss off, so that he wouldn't drive home drunk. The college grad in India wanted to report what he saw. The production assistant wanted to throw her drink in the director's face.

And in each case, these people chose compliance. The bartender poured as small a cocktail as she thought she could get away with. The new grad kept his mouth shut, hoping that someone with more experience would blow the whistle on his colleague. The production assistant ordered a salad and pretended not to hear the inappropriate remarks.

These people knew their values. They knew what their True No would be. But the costs of enacting them were too high—they would lose jobs they desperately needed, or possibly expose themselves to consequences far worse in retaliation. They also believed that in those particular situations, their defiance would not lead to substantive change, and the only person affected—negatively—would be them.

I know the feeling well. My crocodile smile was often an outward manifestation of my conscious compliance. Such compliance has long been a survival strategy, a way to survive in a world in which defiance is often difficult or not always possible. It often looks like an effortlessly placid expression, or an intentional blankness. All of it translates to: *I don't agree with this. I would like to defy in this situation. But I don't believe I can, at least right now.*

Defiance exists on a spectrum, and the road to it can be long. A lot of space exists between unthinking compliance and public defiance, in large part because the costs of defiance are often so high. Many who would like to defy cannot do so—not because they lack capacity, knowledge, or understanding, but because they want to keep their job, protect their family, or even stay alive.

Defiance can be life-changing, but it can also be life-endangering.

Rosa Parks may be revered in the history books now, but she suffered for decades after her refusal to move for a white bus passenger. She was mocked and scorned by those who opposed equal rights, and she received death threats for years. Both she and her husband, Raymond, lost their jobs, and despite her new status as a leading figure in the Civil Rights Movement, she felt neglected and underpaid by its leading organizations. Fearful of the threats on Rosa's life, Raymond struggled with anxiety and began drinking heavily, eventually suffering a nervous breakdown. Rosa developed a heart condition, stomach ulcers, and chronic insomnia from the stress of the constant threats on her life and her family's precarious financial situation. Isolated and broke, the Parkses eventually moved to Detroit, where Rosa endured almost a decade of underemployment before landing a steady job in the office of John Conyers, an African American congressperson.

These costs do not often come into the discussion of Rosa Parks's decision to defy. I mention them not to dissuade anyone from defiance, but to offer a picture of its full human impact. The decision to defy does not make someone an invincible superhero, a breed apart from us; it does not insulate them from harsh and negative consequences. That even Rosa Parks suffered, in the years after her refusal to move, only emphasizes a central fact: She was a human being, one who looked for work and had trouble with her bills and whose marriage sometimes buckled under the pressure of her newfound fame. She could not have foreseen the exact particulars of these costs, but she understood the risks and consequences of her refusal to move on the bus that day—and she did it anyway.

"Yes, I'd do it again," she told a reporter in 1965. "We should all be treated as human beings and not humiliated or be made to feel inferior because of [skin] color, over which we have no control."

Saying "no" on the bus was part of her defiance calculus, the decision-making process she used to determine when and where to act. Defiance was worth its costs for Rosa Parks that day in Montgomery. Every day prior to that one was part of her necessary preparation, a stepping stone along the way to her defiance. When Rosa's grandmother told her that she needed to change her behavior or she'd be lynched before she was twenty, she was advising conscious compliance—telling her to smile at the indignities, to make the racists like her or at least ignore her, so that she could survive one more day. I believe Rosa came to understand this advice not as a message of resignation, but of possibility: If she survived long enough, she could change the world.

Most of us will encounter many situations when the time to endure the consequences of defiance is "not now." We consciously comply because the ramifications of defiance—ostracization, loneliness, financial ruin, threats to our physical safety—are very real. And when we practice conscious compliance, we need to do so without judgment, shame, or self-flagellation.

Sometimes, we simply need to *defer defiance* to another day.

"The place to defy is later," Kenneth told me. "Not on the side of the road with an officer with a gun, but maybe in the relative safety of the courts." This approach allowed Kenneth to stomach the encounters that he felt were unjust and obediently comply—he could challenge the charges later, in the presence of witnesses, instead of alone. Even then, however, the odds might be stacked against him. Kenneth did go to court to appeal his California stop ticket and failed—but at least he only lost a minor court case and not his life.

Conscious compliance requires knowing that our values—the foundation for the beliefs we hold most dear—cannot always come first and foremost each and every day. We note our tension, we acknowledge our desire to defy. But we consciously stop short of escalating our wish to defy into action. If we defied whenever a situation did not perfectly align with our values, we might not survive for very long or even get much done. When defiance is too dangerous or ineffective, or is beyond our capabilities, we're not necessarily rejecting defiance—just biding our time and saving our energy. It is not an excuse but a prelude, one that too often goes unnoticed and unremarked upon.

A True No is often newsworthy: Nelson Mandela leading a crowd of people through Johannesburg, in open defiance of the apartheid state's curfews that forbade the presence of Black people; Mahatma Gandhi leading thousands to the coast of the Arabian Sea, protesting England's salt tax; Rosa Parks refusing to give up her seat in opposition to Jim Crow "separate but equal" laws.

But conscious compliance doesn't usually make headlines.

I would argue that every act of defiance is preceded by dozens, hundreds, perhaps even thousands of moments of conscious compliance: moments when defiance is not declined, but merely deferred.

How many times did Mandela close the door to his home an hour before the curfew, not to open it again until morning?

How many times did Gandhi's followers pay the salt tax before picking up handfuls of salt from the shore in Dandi?

And how many times had Rosa Parks complied with segregation laws before that momentous day in 1955 in Montgomery?

Ordinary people consciously comply every day, planning for the day they won't anymore. Conscious compliance might feel like a swallowed retort to a racist joke, or a clenched fist invisible beneath a boardroom table after a sexist remark. It might not look like anything at all, from the outside. But behind the bitten tongue and the placid expression of a crocodile smile, a True No often stands ready, waiting for its moment.

Conscious compliance allows us to stay on the beam, to keep balancing until we reach safer ground. It is not a decision about whether or not defiance is something we support—it is a decision about what can keep us safe in that moment. But it is a short-term strategy, one that also has a price.

The Costs of Conscious Compliance

I asked Julian, the brother of one of my students and a sociologist whose research interests focus on race, housing, and policing, whether he remembered his first encounter with a police officer. He did, he told me, but it was so traumatic that he didn't want to discuss it.

"It is my first memory," he said.

When I asked him what he would recommend to other Black people when encountering a police officer in a traffic stop, Julian paused.

"I know what we're told to do," he said. "And for me, it depends on the situation. If it's someone I care about, I might tell them what I was told: Bow your head; do whatever you need to do to get out alive."

He paused again.

"But telling someone to do that—it's a lot to ask of someone. If you were in an abusive relationship, could I really tell you to just bow your head for one more day, one more hour, because bad things might happen to you if you don't? The truth is, for Black people in this country, bad things are already happening to you."

There is a cost—emotional, spiritual, psychological, and physical—for this kind of conscious compliance. The effects of repeated conscious compliance compound—that is the price of continually being violated and disrespected over and over again.

"Every day, being beaten down by a world that hates you, and then having to bow your head to make sure you survive—that takes a toll," he said with a sigh.

Again and again, speaking to people about conscious compliance, I heard about the costs. Katya described a sense of inertia at her work, a feeling that she was letting things go, not caring anymore. Neil spoke about feeling dehumanized during his interactions with the police. In my experience with the senior professor, I felt powerless and muted, swallowing my feelings at each objectifying remark.

Conscious compliance helped each of us manage or exit a situation, but those experiences stay with us. It also does nothing to change the larger, problematic environment. To be clear, the onus should *not* be on those who are discriminated against to change the culture. That responsibility should fall to those who are part of the system that administers the inequity. The reality, however, is that it is often the victimized who transform society with their defiance, despite the risks.

In my conversations with Neil, he used a phrase I found powerful: *costly dichotomy*. It refers not only to the cost of defiance but also the often unacknowledged cost of compliance.

Increasingly, Neil found himself triggered by the images of police brutality he saw from around the country, directed at Black men like him.

"You want naturally to save yourself and save your family," he explained. "But then you're like, 'No, I refuse, something is not right. This is wrong.' You want to speak out against it."

Neil's anger was followed by deep sadness. He knew strategies to help him survive situations with the police, but what he didn't know was how to survive his survival strategies. What could he do to assuage this fear, to process this anger, to help overcome his sadness, to right this wrong?

Neil likened his feelings—and the feelings of so many others like him—to the stages of grief. He and so many other Black people he knew were grieving, and they needed to find ways to process that grief: to move through its stages, past denial, anger, bargaining, and depression into something productive.

Neil spent hours reading up on the laws regulating arrests and detainment. He and his sisters regularly hit the streets for what they called "cop watches," monitoring where the police in their neighborhood were patrolling, filming any interactions, and counseling anyone stopped by the police about their legal rights.

Neil knows now what the police can and cannot do, and increasingly he has come to understand that in addition to group action, there are other ways to channel his grief in productive ways.

"You cannot beat the system," he told me, "just going up against it head-on. It's not going to work. This [racism] has been around for a very, very long time. . . . If you think by kicking and screaming and burning down the whole entire city, that that's necessarily going to work in your favor, it is not. Because you've not burned down their city, you burned down *your* city. So, you realize, like, okay. Let me go to law school."

Conscious compliance helped Neil survive individual situations, but over the long term, he had to learn to harness the grief it created. The question became what to do with his grief, how to use it in service of his core moral values, with action that both created potential for positive change and kept him safe.

Neil found his solution in learning the law and community action. Julian, too, recognizes the limitations of the system and the barriers to change. He expressed to me that his focus was not on confronting or attempting to reform the police force, but on helping his own community process the trauma inflicted upon them.

Conscious compliance can keep you safer in the moment, but it also wears on you—physically, mentally, and emotionally. Not being able to react to racism or sexism, overt or otherwise, does more than cause mental distress—repeatedly squashing your emotions can make a person physically sick.

That does not mean that conscious compliance is not often necessary or useful—indeed, in many cases, conscious compliance sets the stage for eventual acts of defiance. But for conscious compliance to serve its true purpose, we must learn to recognize when it is no longer useful to us and how to break out of its grip.

9

Quiet Defiance

atthew was deployed to Iraq when he was nineteen. When I
spoke with him one Sunday afternoon, his voice stopped and
shook several times as he told me his story. His father, a cook,
had been working in a restaurant near the World Trade Center on September 11, 2001, and Matthew grew up with the image of the towers,
smoke and fire pouring from them, seared into his memory. He felt a
patriotic duty to defend his country, and also, as the son of Asian immigrants, to prove that he belonged in America.

At first, Matthew believed wholeheartedly in the justice of the war in
Iraq and the actions of his fellow soldiers in the Marine Corps infantry.
He had been trained to follow orders, drilled into unflinching compliance.

"Being young," he said, "you're vulnerable, you listen to what senior
officers say. In the military, it's yes, yes, yes."

Stationed north of the Euphrates, near Fallujah, Matthew's unit was
often under fire, and a lot of his friends were "getting blown up"—injured
or killed as their vehicles sustained debilitating blasts from improvised

explosive devices (IEDs). This sustained sense of danger and personal loss, he says, intensified the unit's sense of brotherhood and camaraderie, making it even easier to follow orders. "Even if you felt some inkling of hesitation," he told me, "you did it anyway, because you wanted revenge."

A few months into his first deployment, Matthew's best friend was killed. Soon after, he and his squad were returning from a nighttime sweep of an area in the desert, their trucks aligned in formation. Although they were on high alert, they were not that worried about an attack—usually nothing happened at night. But just a few miles from base, near a mosque, they suddenly saw the telltale flash followed by the boom of a rocket-propelled grenade that came too close for comfort.

"Bravo Team! Dismount! Engage!"

Instantly, his squad went into action. Marines jumped from the lead vehicles, running toward the mosque while the other soldiers, Matthew among them, provided cover. Within minutes, Matthew's team leader had returned with four young men. They were dusty and bloody, blindfolded and handcuffed. The other Marines put them in separate trucks. Matthew's prisoner was also yelling at the top of his lungs in Arabic, which Matthew could not understand. Neither, it seemed, could his team leader, an Italian American sergeant from Long Island who was only a year older than him.

"Shut him up," he told Matthew, as the sound of vehicle engines filled the air, along with shouts from his fellow Marines and the screams of villagers.

"What do you mean? What do you want me to do?"

The sergeant squinted at him and tilted his head.

"It's not complicated. Hit him in the mouth."

Matthew stared at the man in front of him, who had blood streaming from his nose and into his beard.

Matthew did not want to hit this man.

But he had a direct order. His best friend was dead. He did not want to disobey his superior or look like a coward.

So Matthew raised his fist and punched him in the face.

"He's still talking," Matthew's team leader said. "Hit him again. Hit him in the stomach."

"And that's what I did."

A few weeks later, Matthew came to learn that the alleged insurgents they had captured were actually teenage boys, around fifteen or sixteen years old, and unaffiliated with any local militant groups.

"I trusted my team leader that was on the ground that these were insurgents. Now I think about whether I did the right thing. I don't know. I didn't question at all."

A little after that, Matthew and his squad were on patrol when they came under heavy fire from AK-47s. Matthew was in the lead truck, listening to the bullets whizz over his vehicle as the voice of a superior officer cut through the static of the radio.

"All right. Bravo dismount, engage north side at this area."

The normal response to such an attack would be to set up a perimeter and return fire in a purposeful way.

"The military's rules of engagement dictate that soldiers must have a clear line of sight and target," Matthew told me. "You can't just 'spray and pray'—that can cause a lot of casualties, especially to innocent people."

But "spray and pray" was exactly what Matthew's unit did. The entire convoy of four trucks rumbled to a stop, set up in a semicircular formation, and aimed their weapons in the general direction of the north. The air was full of the acrid smell of rapidly firing M16s and M4 carbine rifles, and Matthew's ears rang from the repeated reports of the convoy's machine guns. They did not have visual confirmation of their target, but they were unloading their weapons, firing indiscriminately into the glaring white sands of the desert.

Marines are taught to react quickly, to follow orders without hesitation. But Matthew, in that moment, did something different: He turned around. He did not fire his weapon.

His fellow soldiers, he thought, were shooting just to shoot—after months of deployment, the stress of daily patrols, the emotional turmoil of losing friends, they were feeling trigger-happy. The shooting was a release. But it didn't seem right to him.

Within a few minutes, the shooting was dying down. Matthew's team leader approached him, a perplexed look on his face.

"What are you doing?"

"I'm providing security, sir," Matthew said. Providing security or cover for the rear flank was the only excuse he could think of to explain why he was facing the desert behind him and not directly disobeying an order.

"I didn't order that," his superior said. But he was too busy to reprimand Matthew—and what he'd been doing was, in fact, a good idea. So his superior let it go.

This was as close to defying an order as Matthew had ever come. He describes his actions that day as partly instinctual and partly considered.

"That was the first time I actively decided," he said of his choice not to follow an order. "But it was only half conscious. It all happened in less than thirty seconds."

Moving forward—and especially in his next deployment, when he became a team leader himself—Matthew grew to question some of what he saw on the battlefield and in the war more generally. He believed that the best way to balance his duties as a soldier with his values would be to work in intelligence. Remembering the teenage boy he had punched, screaming in a language he did not understand, Matthew began learning Arabic, studying the language with the military's translators. He also vowed to not participate in firefights without following the military's rules of engagement—a clear line of sight of the opposing fire and a target—and always volunteered to ride in the first car of a convoy, where his primary objective would be to scout for IEDs.

His primary mission, he told himself, was not to lead his men into indiscriminate battle, but to keep them safe.

As noble as these plans sound, he kept them to himself.

"Even though I felt all of this, I did not have the courage to say it aloud to anyone else—I was sure they would send me to the brig," he told me, referring to the military prison.

He paused.

"And they'd be right."

Matthew would not define his decision to provide cover as defiant. He described it to me as cowardly, saying that he "didn't have the courage to say it aloud."

But the moment was surely a solution to his increasingly complicated feelings about the war and his role in it. Throughout the remainder of his tour, and into the next, he would employ a similar strategy to avoid situations in which he felt he would be ordered to do things he found unethical. And he grew just a little more comfortable with the idea of questioning his superiors.

In his second tour of duty, Matthew's commander ordered him to lead a nighttime patrol on a high-incline terrain. To do so, the Marines would have to wear night vision goggles, greatly affecting their depth perception. A problem, as the route led directly along a cliff edge, with no margin for error.

After his briefing, Matthew approached his sergeant—his superior in the chain of command—and tentatively told him that he did not think the patrol was a good idea—that it was dangerous and would put Marines at unnecessary risk of wrecking their vehicles.

"Shut up," the sergeant responded. "Do as he says."

"I can't, sir," Matthew responded.

"Are you disobeying?"

"No. I am not disobeying, but I am adamantly against this idea."

"Shut up," the staff sergeant said again. "Get out."

The mission that night was unsuccessful—the soldiers returned to base after thirty minutes, the terrain deemed too treacherous to proceed farther. And although Matthew had been correct in his assessment, he was not thanked. Quite the opposite. A few weeks later, he was denied a promotion, by the very sergeant with whom he had registered his concerns.

Given all this, I asked, would Matthew make the same decision again?

"One hundred percent," he told me. "My concern was not so much with fighting the insurgents. My concern was making sure that each of us came back home alive. It didn't happen in the first deployment, so I was making sure it happened on the second. The only thing I would have changed was having a better argument, providing better evidence . . . or having a better approach.

"But I still would have questioned," he assured me. "I still would have made my stand."

Quiet Defiance

Matthew's behavior shows the way that conscious compliance often predates more defiant behavior. Over the course of his deployments, Matthew learned to recognize his tension and vocalize his discomfort, even though he never escalated into his full-blown True No.

But his story reveals another kind of behavior, which I call *quiet defiance*. When Matthew provided security instead of firing his weapon, he was technically disobeying an order—but he was doing so in a way that could be explained as compliant with a separate and equally important directive. He was not firing at the enemy, no—but he *was* protecting his fellow service members.

Matthew was doing whatever he could to stay true to his higher principles and values, without publicly defying. The costs of defiance, in that situation, might be life or death. But so were the costs of compliance.

Quiet defiance is defiance without the vocal public declaration and without the use of the word "no." It might be done secretly, but it is a deliberate action, like Matthew's, that goes against an order or a rule, without drawing too much attention to the defiant act. It is a way of breaking out of conscious compliance—of taking back our True No.

This kind of defiance is common in regimented or repressive environments. It is a favorite tool of artists and intellectuals operating under state censorship, who have often couched their defiance to repressive regimes in irony, appearing to comply with the censor while subtly evading its grasp.

In one version of Milgram's experiments, the "experimenter," the actor who ordered the participant to give shocks to the "learner," left the room after giving initial instructions. He gave the remaining orders to shock the learner by telephone, thus reducing his surveillance of the participant. Obedience to give shocks dropped from 65 percent to approximately 20 percent, but interestingly, the participants often did not inform the experimenter of their deviation from the procedure. Indeed, some assured the experimenter they were raising the shock level as requested while, in reality, they were repeatedly using the lowest shock on the board. This quiet defiance allowed them to defy the experimenter's orders without conflict.

My quiet defiance with the senior professor who constantly mentioned my exotic race meant working with him as little as possible and avoiding interactions with him whenever I could. In so doing, I was quite literally avoiding conflict, but I was also working toward my goal: to loosen the control he wanted over me. Quiet defiance was a way to honor my will to defy without completely jeopardizing my career.

In the military, where open defiance is grounds for a court-martial, quiet defiance can be an important tool for service members, allowing them to maintain some of their values without sacrificing their livelihoods. That was certainly true for Matthew.

The culture of the Marine Corps is especially driven by hierarchy and obedience to authority. Another student of mine, Jody, who was also a former officer in the Marine Corps, described to me the process recruits undergo during training as going "robotic"—after weeks of drills, screaming sergeants, and reading the 280-page book on apparel and presentation, newly minted Marines began to behave automatically and uniformly. Such behaviors are intentional, part of the Corps's mission to act in a co-ordinated, united, cooperative fashion. Conformity is the goal. The idea is that officers may have to do a lot of difficult tasks, but thinking is not one of them.

"Your job is not to think," Jody told me. "Your job is to do."

Such widespread compliance is efficient, and no doubt can be quite effective on the battlefield. One of the values that Jody most admired during and after his service was the idea that a Marine would do anything for another Marine. The joke, he told me, was that if a Marine gave a group of other Marines a pack of cotton swabs and told them to move a building, they would—no questions asked.

Such unthinking loyalty to the group has obvious risks. Environments like the Marine Corps leave little room for anything but compliance. Quiet defiance like Matthew's is sometimes an option, but often it takes years for people within environments like this to attain the necessary rank to feel secure enacting it. Most exit the military, and environments like it, without ever having voiced their True No. Others find their will to resist fades under constant pressure. Quiet defiance can become quieter and quieter, until it is nearly silent. Conscious compliance, over time, dims into unthinking obedience. And the very people who could change things for the better—if they were empowered and protected—are either silenced or gone.

Not Another Brick in the Wall

Late one night, soon after completing his training, Kevin, a young Black police officer from California, and his partner responded to a call about a bicycle robbery. Other officers were already on the scene, standing in front of a house with an open garage door, he told me. They wanted to enter and search it.

But Kevin had recently been trained in this exact scenario. He knew that to enter a home, even just the garage, the police needed verbal permission, a search warrant, or exigency: someone screaming or the presence of blood.

"We can't just go in," he told his fellow officers, most of whom had twenty years of experience over him.

The street was quiet, only the buzzing sound of streetlights and the distant whine of traffic breaking the silence.

"You don't know what you're talking about, rookie," they retorted.

"I do," he said. "We can't go in unless we talk to the homeowner."

So Kevin knocked on the thin wooden door. He knocked and knocked, but no one answered. Then he spoke to a neighbor, who confirmed that the homeowner was inside—his car was in the driveway. He was probably just asleep, the man told Kevin. After all, it was nearly four o'clock in the morning.

By this time, Kevin's colleagues were even more impatient.

"They were like, 'Eff it,'" Kevin told me. "'Let's just go in.'"

Kevin refused. So did one other officer. The two of them stood by a squad car on the street, watching as their senior colleagues filed into the darkness of the garage, their flashlight beams strobing into the corners.

"In my mind, I was thinking about how things could go wrong," Kevin told me. "How the owner could wake up and come down with his gun and start shooting. I didn't want to be part of that. Because if something bad happens, it's gonna happen to all of us."

Kevin finished his training during the summer of 2020, and the actions of the police officers toward George Floyd were prominent in his mind. This incident as well as his family and his mentor, a retired police captain, had taught him not to simply obey without question.

In the end, the search resolved without incident. There was no evidence of theft in the garage, and when the homeowner woke up, he thanked the officers for checking in on his property.

But it was a long, quiet ride back to the station for Kevin and his partner. He knew he had done the right thing, but he also knew what it would look like to the others. As a young cop still on training probation, publicly defying senior officers was a risky move.

Kevin felt the consequences almost immediately after his return to the station: long stares in his direction, whispers behind his back, rooms going silent when he walked in. Within an hour, he was called into the sergeant's office.

"Read this," the sergeant said, closing the door and handing Kevin a stack of papers.

The papers, he claimed, gave legal permission to search without a warrant.

Kevin says that he is comfortable admitting when he is wrong. But in this case, he knew he wasn't. Not only did he remember his training, he knew the law. And although he was shaking, his hands trembling as he perused the papers in his sergeant's office, he stood firm. He knew the papers did not mean what the sergeant said they meant.

"With all due respect, sir," he repeated, "we had no standing."

This incensed the sergeant.

"How are you, a twenty-something rookie, going to tell me, a veteran, how to do my job?" he fumed.

Kevin didn't back down. He hadn't joined the force to be a follower, he told himself. He wanted to do the job right. So he insisted, again, that he had simply followed his training, and that what he had done was in accordance with both his instruction and the law.

"I knew before joining that this job would be tough. I thought about what I wanted to say before I went into the sergeant's office. I took a deep breath before speaking up, because sometimes I do get flustered. I do stutter and get nervous. But, you know, I always want to do the right thing."

For this, Kevin was ostracized, reprimanded, and eventually transferred to a different rotation. He was no longer trusted.

"This one incident went around the whole station."

It didn't matter that he was right. What his sergeant and his fellow officers were objecting to was that he had said no.

Speaking Up

The hierarchy of defiance is strong, but conscious compliance is not only the purview of the nondominant class. Often those with real power and authority, including police officers themselves, are in cultures similar to ones Katya and I experienced, which strongly discourage defiance.

The "blue wall of silence" refers to police officers' refusal to tell the truth if that truth implicates a fellow officer in wrongdoing or would besmirch the reputation of the force. It is pervasive, encouraging good police officers to say nothing to prevent the bad behavior of other officers. In a 2000 survey presented at the Annual Conference of the International Association of Chiefs of Police, almost half of the respondents—46 percent—

claimed to have witnessed misconduct by another police officer, and to have concealed what they saw. Their reasons varied but often hinged on job security and personal safety. Officers rely on one another for survival. The officers also believed they would be ostracized, fired, or blackballed, or that their reporting would not make any difference.

As law enforcement veteran Michael Quinn writes in his book *Walking with the Devil: The Police Code of Silence,* the blue wall often corrupts honest cops. It allows the police to take the law into their own hands because they know their peers will not "snitch" on them. Often honest cops have no one to turn to because even the higher-ranking officers are involved in acts of wrongdoing. The code of silence inevitably accumulates into unethical, criminal behavior among some officers and, as Quinn writes, changes the police motto "protect and serve" to "convict and incarcerate."

Quinn was a police officer in Minnesota from 1975 to 1999, and he confronted the blue wall again and again during his career. Once, several years into his time on the Minneapolis PD, he encountered a badly beaten sex worker near a dumpster. She named two police officers as her assailants, but declined to file a report, because they had threatened to kill her if she snitched on them. Quinn's partner followed the code of silence—if the sex worker did not want to report who had assaulted her, he would decline to pursue it.

At first, Quinn did the same. The two officers in question were notorious for their brutality and corruption, and Quinn himself had recently been ostracized for reporting a colleague's drinking on the job. He hadn't enjoyed being treated as a pariah. So, following his partner's lead, Quinn decided to "walk with the devil": to let this obviously illegal assault slide, even though he knew it was wrong. He would keep silent to stay in line with the code.

But his conscience got the better of him, and over dinner with his lieutenant and the deputy chief of police a few days later, he told the whole story. Word got out, and the two officers who had beaten the sex worker cornered Quinn in the Hennepin County courthouse, shoving him against the wall and threatening his life in much the same way they had done with the woman they had beaten:

"If you snitch us off again, we'll kill you," they snarled, close enough that Quinn could smell their breath.

Quinn exited this encounter badly shaken and frightened for his life. But although the two abusive officers were not held accountable, he noticed that in the weeks and months after, there were no more sex workers battered in his district. For Quinn, this was evidence that the code of silence can be broken—all it takes is one person to speak up. When he broke the code, misconduct around him stopped—other police officers knew that he would report wrongdoing if he saw it, so they did not engage in it while he was around. His act of defiance changed the water in which the other officers were swimming. If more officers did the same, Quinn argues, police misconduct would decrease across the department.

As Quinn writes, the code of silence is absolutely corruptive. It harms police officers, who find themselves torn between their ethical responsibilities and their safety. It harms civilians, who are more likely to encounter police brutality or injustice. And it harms the police's reputation and standing within a community, destroying any trust between law enforcement and the people they are supposed to protect.

But to Quinn, the blue wall of silence is not impregnable. He has spent the past twenty years, since his retirement, educating officers, departments, and the public on how it can be toppled. As his story shows, sometimes all it takes is one person to bring much needed change.

Escalating Toward Defiance

As police officers, Michael Quinn and Kevin occupy very different places on the power structure than Black civilians like Neil, Kenneth, or Julian. But like them, both Quinn and Kevin know that costs are an inevitable consequence of both defiance and compliance. They know that a True No brings with it many outcomes, some of them potentially harmful to themselves. Both had to decide when conscious compliance was their best option and when it was time to defy.

This can be a difficult decision, heavily dependent on the situation and our own capacity for risk. But I have found a helpful framework for making this decision. The literature on medical activism offers a useful road map for deciding how to proceed in situations where simply following the established procedure is not enough.

Medical activism describes the ways that healthcare workers can advocate for their patients. It encompasses several broad, escalating defiance behaviors beginning with *routine advocacy* and culminating in a *principled exit*. I have adapted some of the basic principles of medical activism into a structure for broader use. And you don't need to be a doctor for this four-part framework to be useful (see figure 5).

Advocacy → Dissent → Disobedience → Principled Exit

Figure 5: Escalating Defiance in Institutions

Routine advocacy means advocating for your values during your regular daily activity. Doctors do this, for example, when they fight for a patient's necessary medical test to be covered by a reluctant insurance company. Michael Quinn did this, initially, when he encouraged the beaten sex worker to report the crime against her, but he failed in convincing her. Matthew, the Marine, used a form of advocacy, quiet defiance, to protect his comrades and not violate his values.

If advocacy fails, you can move to internal *dissent*—knowing that it can sometimes raise tensions or backfire. Matthew engaged in internal dissent when he privately approached his sergeant on his second tour and voiced his concerns about the unsafe mission. If internal dissent does not work, public dissent—another way of escalating and vocalizing tension—is usually next in line. Jeffrey Wigand's appearance on *60 Minutes*, detailing the ways in which Brown & Williamson manipulated its ingredients, was an act of dissent that greatly increased the public's knowledge concerning Big Tobacco's deception, while also exposing Wigand to increased scrutiny and harassment.

Professional disobedience is the third stage and involves publicly disobeying rules or laws that are causing clear harm. For a doctor, this might mean refusing to comply with a law refusing medical service to a particular class of people—undocumented immigrants, for example. One woman I spoke with, a former grants coordinator named Sara, discovered serious ethical violations at a healthcare nonprofit where she worked. After her internal dissent failed, and she was concerned that public dissent might lead to interrupted care for vulnerable cancer patients, she began to engage in professional disobedience, refusing to attend meetings if doing so would embroil her in what she saw as clear corruption. It worked, for a while.

But ultimately, it led her to a *principled exit,* which often arises after the other stages of activism have been ineffective or would cause too much harm—either to yourself or to others.

Sara recalled sitting in her boss's office for the final time, being reprimanded for an instance of internal dissent. When she thought about complying one more time with behavior she found unethical, she found herself thinking, *This isn't me.* It was a powerful realization.

Principled exits often bring with them psychological benefits like this: decreased stress, lowered anxiety, and a newfound relief from the daily psychological struggle of conscious compliance and quiet defiance. Such exits are not always a first course of action, and they are not always a viable possibility. But they can be a way to preserve your sense of self when all other methods at effecting change have failed.

Escalating from routine advocacy to dissent, or from dissent to disobedience to exit, can provide more possibilities for the expression of defiance, minimizing the costs of leaping straight to whistleblowing or resignation. It may also lead to gradual culture change within environments in real need of reform.

However, two whistleblowers I spoke with gave some words of caution.

Jack, a doctor, reported avoidable deaths at his hospital that were caused by the incompetence of his colleagues. Denise, a consultant investigating sexual harassment in the workplace, reported an inept Human Resources department. Both of them lost their jobs after blowing the whistle, and experienced significant financial, mental, and health costs. At one point, Jack contemplated taking his life.

Yet when I asked, "Would you do it again?" Jack replied emphatically, "Yes. I would do it again. But . . . perhaps not on my own."

Denise's response was more measured at first.

"The costs were just too high, so I'm not sure I'd go about it in the same way," she told me. After some reflection, she added, "But would I speak up again? Absolutely. Now I know better, I would go directly to the press rather than trying to handle things internally."

Skipping internal reporting and proceeding directly to public dissent, followed swiftly by a principled exit, would have mitigated some of the retaliation she tolerated, Denise told me.

Not every situation is as drastic as these, and changing an institutional culture is sometimes possible. I found evidence of this in my interviews with current police officers. Along with those who noted the cultures of extreme obedience and compliance, a few officers, including Kevin, referred to an era of "new policing," in which the old ways of lockstep conformity were slowly eroding under the pressure of outside scrutiny and a new crop of socially conscious officers.

After his transfer, Kevin was fortunate to join a team with a much wider knowledge of the law and a more open, transparent approach to policing that aligned with his values. He feels that younger officers, like him, are more inclined to work in ways where the old ways of policing— which he describes as silent, retributive, and secretive—don't hold sway.

No one can tell you when it is the right time for you to defy or to comply, but if we can become more aware of our decisions, while understanding the costs and benefits of any choice we make, we can preserve both our safety and our sense of self. We can use conscious compliance or quiet defiance to buy time, so that we can mitigate the risks of public defiance, maximize its effectiveness, and understand when the scales have tipped—when we can no longer quietly comply with a situation that goes against our sense of who we are and what matters to us.

Conscious compliance and quiet defiance require awareness: of yourself, of your values, and of your surroundings. They mean knowing when to say yes, but also knowing when to stop. Because while not all conscious compliance ultimately leads to defiance, it is often a harbinger of things to come: an indication that although now might not be the right time to defy, that time may be fast approaching.

Conscious compliance can become conscientious rebellion, and quiet defiance can become loud very quickly.

Part Three

BECOMING A MORAL MAVERICK

10

Who Am I?

met Pradnya when she was in her late forties. An immigrant from India who had been in the United Kingdom for about twenty years, she was a petite woman with neatly parted black hair that hung past her shoulders, wearing a dark blue sari under a heavy down coat. She was visiting her general practitioner in Edinburgh with complaints of fatigue, nerve pain, and other systemic bodily ailments.

As a medical student, it was my job to greet her first and take her medical history. When I examined her file in preparation, I was surprised to read that Pradnya had undergone a hysterectomy in her early twenties in India. Now, sitting across from her in the small medical office, I asked her why.

"So I wouldn't have any more children," she replied.

In a quiet, halting voice, Pradnya told me that she had been married at age thirteen and had her first child by age fifteen. After her fifth child, Pradnya's husband arranged for her to have a hysterectomy at the age of twenty-four so that she could not conceive again. She received no follow-up care for her enforced early menopause.

After Pradnya moved to Scotland, National Health Service doctors ignored her ailments. Perhaps it didn't help that Pradnya did not speak English fluently. She could communicate, but she was not particularly detailed in describing her symptoms. Indeed, after Pradnya left the office, her doctor complained to me about her lack of fluency and told me that he had never truly believed her complaints or the severity of her pain.

I kept my face blank when he told me this, clutching her file, but inside I was devastated. Pradnya's care had been shockingly inadequate in India, and she had subsequently been neglected and abandoned by her U.K. doctors, too. Later that day in the break room, I heard the doctors banter over their afternoon tea—making fun of their "difficult" patients, the "scoliosis" or the "kidney stones," referring to them not by their names, but by their ailments. I walked out, unable to stay or to say anything.

I couldn't sleep that night, gutted by how the medical establishment had utterly failed Pradnya. Tossing and turning, I kept seeing Pradnya's face—her sad brown eyes, her hair neatly parted and plaited, reminded me of my own mother. Staring into the darkness, I realized that what bothered me so much about Pradnya's visit was that she didn't even seem to expect anyone to help her anymore. She had grown used to being ignored, her concerns minimized. Coming from a family of Indian immigrants, I had firsthand experience of our family not being taken seriously, of being dismissed and discriminated against. Perhaps that was what engraved this patient in my mind, memory, and heart: It felt close to home. I knew people like Pradnya, and I knew how they felt. I also realized that the Hippocratic Oath—*first, do no harm*—was not enough to help people like her. As doctors, we had to do *more*.

That short experience with Pradnya helped me become aware that doctors are not neutral observers of social inequalities; if they do nothing, they become part of the system that reinforces injustice. Over the next few years, I began to understand how a person's autonomy can affect their health and well-being. I learned that higher incomes and education levels were often better predictors of patient well-being than the availability of healthcare services. I saw every day how our health was heavily determined by social factors, subject to the same discriminations that affect so many other aspects of our lives.

A few months later, and a world away from the cold gray skies of Edinburgh, I began working in a small clinic in the warm and sunny but remote Cook Islands, in the South Pacific. While supervising an outpatient clinic at a hospital in Rarotonga, I saw patients presenting with late-stage heart failure: severe, pitting edema, with huge swelling and deformity in their legs. I had seen these symptoms of heart failure before, in the U.K. But those were cases we caught early and could effectively treat. In the Cook Islands, patients visited the hospital in the late stages of serious illness, severely compromised by their condition, barely able to walk. A lack of education, income, and other factors had kept them from visiting the hospital, and a deficiency of resources had limited the options available to them when they finally did seek help.

It was sometime in that year that equality and fairness began to solidify as core values for me, guiding not only the way I wanted to practice medicine but as something I wanted for everyone in all areas of their lives. Injustice bothered me viscerally. It became increasingly clear that what I was seeing in my exam rooms—Pradnya's downcast face, or the swollen, indented legs of those men and women in the Cook Islands—were not random medical conditions but the aftereffects of discrimination and inequality. I often exited those rooms with my stomach in knots, despondent not only for my patients but for the system I was a part of. Late at night when I was on call and couldn't sleep, I'd play my CD of The Cure's *Disintegration* and pace the large bare room in the clock tower of the hospital, asking myself what I could do to address the situation—to change things for the better. Why did healthcare help only some people and leave so many others behind?

I turned to senior physicians for answers. I asked surgeons, psychiatrists, pediatricians, and general practitioners: Why did they become physicians? What did they love most about their work? How did they help their patients?

But their reactions were often disappointing. Many of them mentioned only their paycheck as their purpose to practice medicine. I ultimately came to understand that this cynicism in many of my fellow doctors was their release for the stress and trauma they experienced on the job. Many of them were too exhausted and overworked to search for meaning.

But if you caught them on the right day, you might see a glimmer of what animated them in their work: They found deep satisfaction in treating their patients, a pleasure in finding a diagnosis and helping them heal. They knew their work was important; they believed, in short, that they could help people be well.

Throughout my medical career, I clung to these moments, when I saw them in other doctors and when I experienced them myself. I began to realize that medicine, to me, was most effective when it addressed not only a patient's symptoms, but the social realities of their life. Medical history is filled with accounts of forced procedures, patients and sometimes entire communities robbed of their agency. As recently as 2019—and thanks to a whistleblowing nurse named Dawn Wooten—*The New York Times* reported stories of forced surgeries, including hysterectomies of many women at a U.S. Immigration and Customs Enforcement detention center in Georgia.

Stories like this are horrific, but they are indicative of the way many patients are treated in exam rooms across the world. All too easily, patients are pressured into things they do not want or understand. Too many times patients are denied the most basic right that doctors should afford to them, and that we should also afford to each other: the right to say no.

Time and time again, these stories highlight the mistreatment of disadvantaged groups and have implications far broader than medicine. When compliance from groups with less power is taken for granted, and consent automatically assumed, the existing inequalities of our world are reinforced and perpetuated with devastating consequences.

To fulfill their primary responsibilities to patients, physicians have obligations not just to "do no harm" but to promote a just healthcare system that leads to well-being for all. As other physicians have proposed, this includes advocating for the fair and equitable allocation of resources and medical services, and widening our focus and responsibilities beyond the boundaries of the examination room or clinical encounter.

I had been raised to believe that medicine was, in the words of my father, the "best thing I could do." But increasingly, as I grew more and more aware of the constrained system I was operating within, I saw that my true calling lay elsewhere. I wanted to delve deeper into the moral foundations of medical practice, and I wanted to investigate the psychological mechanisms of professionalism, conflicts of interest, and informed consent—enormous, persistent issues in healthcare that affected not only so many of my patients but also so many doctors like me. Clinical medicine had made me question: Who was I really? Was I destined to be a "good doctor" like everyone expected? Beyond the opinions of others— my proud parents, my well-meaning teachers—what really was the best thing *I* could do?

Connecting with Yourself

James G. March, an American political scientist, wrote that we often decide how to behave in a situation by answering three questions:

1. Who am I?
2. What kind of situation is this?
3. What does a person like me do in a situation such as this?

Most of us don't ask ourselves these questions explicitly. But there is tremendous value in taking the time to answer them.

I believe these questions, in practice, are less a sequential list than they are a cycle (see figure 6 on the next page). We do not ask them once, but over and over in the course of our defiance journey. I view the questions as a navigational tool for introspective decision-making, like a compass that guides how and when we defy.

The process often begins with the fundamental question, "Who am I?" which urges us to delve deep inside to identify our core values. From this understanding, we can proceed to assess "What type of situation is this?"

prompting us to evaluate the environment for safety and the positive impact we could make. Completing the cycle, asking, "What does a person like me do in a situation like this?" invites us to align our actions with our responsibilities and values as well as to increase our skills and confidence—our ability—to defy. Our actions consequently affect what we think of ourselves, leading back to the first question. This continuous loop of reflection and action guides us through our decision-making process, ensuring that our choices are not just reactions but reflections of our identity.

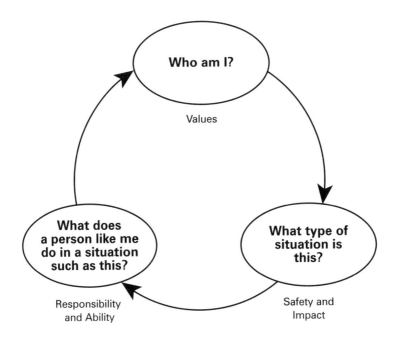

Figure 6: The Defiance Compass

Defiance first requires that we know our values. Whether in a single momentary decision or in more considered choices for one's career or relationships, defiance often goes against habit. It requires conscious self-assessment of what matters most to us beyond our rationalizations, and past our wiring and acculturation.

There are many ways to begin to articulate your values. Some connect with their values through spiritual practice, some through conversations with loved ones and trusted community members. Some connect with their values simply by being in conversation with themselves, deliberating on important events in their lives and reflecting on what those events revealed.

Others discover their true selves by experiencing unfamiliar situations or traveling to other countries. Research shows that people who move abroad are more able to articulate who they are and what is important to them. In a foreign land, you have to grapple with what behaviors are important to you—those that are prominent in your home culture or those that define your host culture. It often starts small with specific behaviors that can grow into a deeper examination of one's core values. For example, if you were raised in a culture that prizes working long hours and productivity, you might question that behavior after moving to a place that gives more emphasis to personal and home life. You need to decide whether high achievement is really a self-defining trait or a culturally driven behavior that you're better off without.

The trigger for such self-examination is not mere novelty, but deep immersion. When it comes to living abroad, studies show that it is not about variety, but rather *depth,* that leads to clarity about oneself. In other words, it's not the number of countries you visit, but the amount of time spent in a particular culture that gives rise to new insights.

For many of us, living through the COVID-19 pandemic was like moving to a foreign country on a one-way ticket. A new way of life was imposed upon us, one in which we were forced to make hard choices and adapt to tough circumstances. I personally found it incredibly challenging to weather the unfamiliar environment of pandemic-land, but it left me with a clearer sense of myself and what I hold dear.

Existing in slight tension with one's environment can be a clarifying experience. But the situation need not be uncomfortable to offer a new perspective. Sometimes simple changes in one's responsibilities—what psychologists call "role entrances," like starting a new job or becoming a parent—can upend our habitual view of the world, and of ourselves, reminding us of who we are, who we wish to be, and what we would like to see in the world.

And when you begin to articulate your values, you may realize that your actions and your stated ideals do not sync up. Which is more accurate, your idea of who you are, or who your actions actually reveal you to be?

Who Do You Think I Am?

A few years ago, as part of a leadership course I was taking, I had to ask my work colleagues and collaborators a question:

"What do *you* think is my greatest passion?"

Asking your colleagues to give you an assessment of your strengths and weaknesses is something of a cliché in business settings. But the question I was asking was different—not *What am I good at?* or *Where could I improve?* but *What do I truly value? What do you think gets me up in the morning?*

Although nerve-racking, this exercise is usually an overwhelmingly positive experience. Not just because people focus on your strengths, your passions, and your values, but because they will tell you things that both validate your values and also reveal views you might not expect.

The exercise produced insightful answers, for example (I hope you'll forgive me if this sounds indulgent):

- If people have a misperception, your mission is to improve things in the world to illuminate that. You have a strong sense of justice. It results in being mission driven with focus. [You] don't care what other people think.

- You care a lot about people behaving justly. You study conflicts of interest due to the fact that you think it alters people's better judgment and makes them act in a less competent and less fair way.

- My sense is that you really enjoy seeing some actual change or impact on the world that improves people's thinking and, through that, improves their welfare. You have the confidence to speak your mind if it's something you're passionate about.

A common theme emerged that resonated well with what I consider my core values: justice, fairness, and the truth. This made sense to me, because my experiences as a child inform my understanding of the world. I saw early on how prejudice, racism, and xenophobia made the world difficult for my family. These early experiences—being shoved in the street, hearing ugly racial epithets, watching my father's career flounder because of his ethnicity—instilled in me a deep yearning for fairness and justice.

Still, another theme emerged from my colleagues' comments that surprised me: that I was unafraid to speak my mind. Becoming comfortable with outward signs of defiance has been a lifelong struggle for me. The same childhood that instilled in me a deep desire for justice and fairness also brought with it the key associations between being compliant and being "good."

It took patients like Pradnya to show me a new way to act in accordance with my values. It took my growing awareness, in childhood and beyond, that authority figures, like teachers, doctors, and the police could sometimes be greatly flawed—petty, uninformed, incompetent, or unfair. When I saw others do things I knew were harmful, I came to understand that achieving fairness and justice sometimes requires opposition. If I wanted to live in a fair and just world, I could not simply coast along. I would have to throw off my ingrained ideas about compliance. I would have to go against the flow.

Learning to defy and speak up took years and is still challenging for me. But my values and principles were instilled in me early on. The lessons from my parents guided an evolving understanding of defiance. I saw the moments in which they were dismissed and discriminated against and felt their sadness. But I also saw how they accepted and welcomed everyone around them with kindness, fairness, and generosity.

Values are not situation dependent, though in practice acting on them may be. Core values do not change if we are at work or at home or dealing with a loved one or a stranger. They are not short-term or tied to specific goals. They come from deep within us, guiding our behavior and framing how we understand the world.

In my exercises with students, common moral values emerge again and again: *integrity, equality, compassion, humanity.* These widely shared core values transcend cultures, religions, contexts, race, and gender, and cannot be boiled down any further. Their apparent simplicity belies their immense power and depth. When we connect with values like these, our actions display our true consent and our true dissent.

Often, our values come from early, formative experiences: tapping into those experiences can guide us toward what matters most. Regina, a student in my healthcare leadership course, described her most important value as "the reduction of suffering." She discovered it at age fifteen, when a close friend was diagnosed with terminal cancer. Watching her friend lose her hair, endure immense pain, and suffer from morphine withdrawal made Regina feel desperate and powerless. After her friend's death, she vowed to do everything she could to help cure cancer. This decision has shaped her choices, her education, and her career. Everything she has done in her life, she realizes now, derives from mourning her friend's death and the despair she felt.

Trevor, another student in the same program, discovered his values as a young child in the Boy Scouts. But it wasn't the Scout Law—which implores boys to be "trustworthy, loyal, helpful, friendly, courteous, kind, obedient, cheerful, thrifty, brave, clean, and reverent"—that taught him his greatest value. It was the failure of his Scout leader—the person charged with teaching that law.

For weeks, Trevor and a dozen other boys at his church had been preparing for a weekend camping trip at the Ten Mile River Scout Camp, planning hikes, buying equipment, and practicing how to build a campfire. On the day of the trip, Trevor was proud to put on his freshly ironed tan uniform. Driving to the church, where the troop's van was loaded up with tents and canned food, he gazed at the empty sleeve of his uniform and imagined the merit badges that would one day fill it. He could not wait to set off for the forest and set up camp.

The mood in the parking lot was buoyant. Everyone was excited for the trip. Trevor and the other boys loaded the van with their equipment, joking and laughing as they waited for their Scoutmaster to arrive.

At first they didn't worry, when the appointed hour came and the Scoutmaster was nowhere to be seen. Maybe he was just running late. Maybe he was stuck in traffic.

But as time went on—thirty minutes, an hour, two hours—they began to grow anxious. Where was their leader?

"Don't worry," their parents told them. "He'll come."

But he didn't come. Three hours passed, then four. The boys stopped joking and horsing around, their smiles turning to disappointed frowns.

After four and a half hours, Trevor and the rest of the boys went home.

"He just didn't turn up," Trevor told me, still in disbelief at what had happened, almost twenty years later. "He just didn't show."

Later, Trevor would hear rumors about the Scoutmaster's drinking. And ten years after this incident, while visiting his hometown church, he would be shocked to hear the Scoutmaster's fond recollections of that weekend camping trip, which he described in detail, even though it had never happened.

For Trevor, this disappointing childhood story didn't illuminate the virtues of timeliness or the dangers of alcohol. It showed him the value and importance of *integrity*—to be honest, straightforward, and to follow through on his commitments. He realized that who the Scouts say they are can be quite different from what they actually do.

"As a child," he said, "you cannot comprehend someone's drinking problem. I vowed I would never be that man who let others down."

Trevor did not want to become the kind of man who strands a group of children in a parking lot, after weeks of promising them a weekend camping trip—or the kind of man who would, after the fact, lie about what had happened. Later, as a doctor, a father, and a husband, he wanted to be dependable and honorable, and to act with integrity in everything he did.

"I think about this value every day of my life," he told me.

So often our values go unvoiced, unexamined, and therefore unavailable to us when we need them most. We often don't think to verbalize our values until we are explicitly asked to list them; they are part of who we are, but because we may not have spent the time to deeply consider them, they can feel inchoate or inaccessible.

For our values to guide our behavior and lead us to a True Yes or True No, we must know them and name them so they can shape our behavior and help us determine how to respond in challenging situations.

I often think of how things might have been different for someone like Pradnya if her doctors had reflected and verbalized their values the way Regina and Trevor did. Many people go through life like Pradnya's doctors: overworked, burned out, and exhausted. It is easy to lose sight of your core values when each day—each hour, each *minute*—brings with it a new challenge, a new storm to navigate. It is easy to simply career from wave to wave, hoping things will calm down.

Reflecting deeply on our core values is something we are rarely encouraged to do, but it is an inherently rewarding and useful exercise. The act of writing out and explaining what our values are and why they're important also makes it more likely that our intended behaviors will follow. There is also evidence suggesting that clarifying our values can lead to a reduced biological stress response, including lower cortisol levels.

Made explicit, our values can guide us, steering us through situations that are complex, difficult, or overwhelming. They can remind us who we are and who we aspire to be. They are the familiar, weathered landmarks that lead us back to ourselves, back to dry land.

This Is Not Me

Several years ago, when I joined a new university as a professor, I remember an incumbent faculty member who withheld vital information required for me to co-teach a course with them. I felt I was being set up to fail without the materials and resources I needed. I did not bring up the issue until months after the course had ended (luckily successfully), when news of it came to the attention of the associate dean, my boss. She was embarrassed, and went to great lengths to reassure me, as a new faculty member, that this was a collaborative institution.

"Sunita, this is not who we are," she insisted.

When I heard that the same incident arose again the following academic year between the same older faculty member and yet another new colleague who had just joined, I thought about my associate dean's words once more.

"It's not who you *want* to be," I thought, "but if it's happening again and again, perhaps it *is* who you are."

Actions speak louder than words, especially when those actions repeat.

Sara, the grants coordinator who finally left her organization, came to a similar realization sitting in her boss's office, after again being told to turn a blind eye to corruption.

This isn't me, she realized before resigning.

So many of the people I have interviewed about defiance over the past few years recall a similar phrase, when they found themselves complying with something they didn't wholly believe in, or when they felt pressured to do something they didn't want to do. The phrase was always something like: *This isn't me. This isn't who I am.*

Such an inner monologue is a manifestation of tension. And it arises from a fundamental moment of cognitive dissonance: the discomfort of a discrepancy between two conflicting beliefs. Who you think you are from your stated values (the first question on the Defiance Compass) is in conflict with the values your actions reflect (from the third question on the Compass). These competing beliefs create pressure that must be resolved.

Here I am, we think, *doing a thing that does not match who I think I am.*

When Jeffrey Wigand took the job with Brown & Williamson, he knew it was an odd fit—a biochemist with a long resume in healthcare, going to work for Big Tobacco. Perhaps he believed he could preserve his values while working to develop a safer cigarette. After a short time, however, he realized that his employers' deceptions about their clearly harmful products did not square with his own sense of self. Integrity was important to him and so was not hurting others. It wasn't him to be complicit in hiding harm to the public. Becoming a whistleblower had costs, but it did more than simply assuage his conscience. It allowed him to live in accordance with his core moral values. Working at a tobacco company clarified, for him, what was truly important, and he has spent the years since speaking out against the lies that his former employers wanted him to propagate.

When our public behavior differs fundamentally from our private beliefs, *who we are* changes. We risk becoming a person we don't recognize, a person we don't believe we are. People gradually start to internalize the norms they publicly display, and our own private opinion often changes to bring itself closer in alignment with our overt behavior, corroding our initial identities. Jeffrey Wigand realized that being a "good employee" at Brown & Williamson meant becoming someone he didn't want to be. The discrepancy between his actions and his sense of self grew wider with every lie about a safer cigarette. If Wigand had remained working quietly and compliantly with his firm, either his sense of integrity would have waned, or he would have had to engage in greater and greater self-deception to continue doing his job and maintaining his sense of self.

We are what we do, regardless of how we see ourselves—and what we do, over time, affects our self-perception. Not to mention the way others perceive us.

A harsh truth that many of us learn is that the realities of life often do not match our ideals. We are not always able to control how the world works. But when we notice that our actions do not match our core values—that our behavior does not reflect who we want to be—that is a signal that something is amiss. When we hear the little voice in our head saying *This isn't me,* it's a good idea to listen.

This is the voice of our values. And that voice should be a wake-up call. It certainly was for Jeffrey Wigand, Sara, and many others. Maybe they had silenced that voice in the past, or maybe it had grown louder and louder as their circumstances changed, rising from a whisper to a roar. But when they finally heard it, increasing in intensity, reminding them of their core values and higher principles, they ultimately took notice. And they defied.

All of us have this voice inside of us, even if we have different relationships to defiance. Some have been comfortable bucking expectations and societal strictures since childhood. Others, like me, were raised to be compliant and obedient, and had to learn, over time, how to act in ways that reflected our true values—including, when necessary, being defiant.

People who defy learn how to listen to their inner voice, so that they can behave in ways that are true to themselves—to cut through their self-deceptions and rationalizations, their entrenched ideas about obedience, and act in accordance with their values.

They don't always do this overnight. Like the pursuit between our independent and interdependent ideal selves, living in total harmony with our values is also an ideal.

As one of my students put it:

Living by my values is an aspiration but never quite a total realization.

We may never reach the fullest realization of our values, but the pursuit imbues our lives with meaning. Self-connection is less of an end goal than a constant learning experience. It's something we can get better at over time. It helps us arrive at our higher principles, our deep and lasting values. It cuts through every self-deception, refuting every story we tell about ourselves. By removing the blinders from our vision, self-connection empowers us to voice our True Yes or our True No.

Defiance starts with one question, short but not so simple:

Who am I?

But it does not end there. To continue toward our True No, we must also direct our gaze outward—to the world we hope to change.

11

The Right Place
and Time

Seen from above, Levi's Stadium, in the San Francisco Bay Area, is a sea of deep green grass, crisscrossed with white hash marks. The red and white logo of the San Francisco 49ers is enormous in the center of the field. In the slanting late afternoon sun, the stands are half empty—there seem to be more players than spectators. Everyone—fans in the stands, players and cheerleaders on the field—is standing.

Except one person. He is tiny in the photo, hard to pick out. You could easily miss him, sitting on a bench on the sidelines, with only his head and shoulders in view. Between two orange Gatorade buckets, the top of the number 7, emblazoned on his crimson uniform, is barely visible. He is facing away from the camera, not yet fully displaying the voluminous Afro that will become his trademark or the fiercely determined gaze that will soon become familiar to people across the globe, even those who do not follow American football.

It is not the first time he has sat during the anthem. He did the same thing during the team's previous two preseason games. But because he was injured and not in uniform, no one seemed to notice.

Tonight is different. He has suited up, pulling the team's regulation gold pants over his surgically repaired knee, fitting bulky shoulder pads over his partially torn shoulder labrum. He has carried his gold helmet onto the field and set it next to him, on the bench. He knows he will not play, but he is ready.

On average, it takes about one minute and fifty-five seconds to sing "The Star-Spangled Banner." For twenty of those roughly 115 seconds, before this trivial preseason game against Green Bay, Colin Kaepernick is still primarily a football player: a young phenom battling his way back from injuries, just three and a half years since leading his team to the Super Bowl.

But then the shutter snaps.

On that night in August 2016, when a local beat reporter posted a zoomed-out photo of the San Francisco 49ers, fans and journalists quickly noticed what had escaped their attention for weeks: Colin Kaepernick, the team's injured quarterback, sitting down during the anthem. At first, his refusal to stand was interpreted as a statement against the military.

But in an interview one day after the photo of him sitting went viral, he made his reasoning clear:

> I am not going to stand up to show pride in a flag for a country that oppresses Black people and people of color. To me, this is bigger than football, and it would be selfish on my part to look the other way.

One month earlier, on July 5, Alton Sterling had been shot six times at point blank range in Baton Rouge, Louisiana. Just one day later, Philando Castile was killed during a routine traffic stop in the Minneapolis–Saint Paul area, in Minnesota. These were just the latest in a long line of Black men and women killed by the police. Kaepernick had been reading about them for years, deeply disturbed by the injustice of their murders, and in recent months, he had been educating himself about systemic racism and nonviolent resistance.

But Colin Kaepernick's education had not been merely between the pages of a book or in the audience of a lecture. He is biracial, the son of a Black father and a white mother who gave him up for adoption soon after he was born. Raised in California by a white family, he had long straddled worlds: Black and white, athlete and student. His whole life, he had attempted to understand who he was, how the world saw him, and what he stood for.

His decision to sit during the national anthem was not to go viral or gain attention, he said at a press conference two days after the photo first gained notice. It was purely a personal decision. It felt wrong to stand up when people were lying dead in the street.

"When there's significant change," he said in a press conference, "and I feel like that flag represents what it's supposed to represent . . . I'll stand."

Kaepernick knew the possible consequences: public outcry, condemnation, possibly the loss of his job. He knew how the media would respond, and he knew that the fan base of the National Football League would be hostile. He anticipated the boos, the slurs, the talking heads on television and radio questioning his patriotism.

He was disrespecting the flag.

He was ungrateful.

He was dishonoring the memory of veterans.

If he didn't like it here, why didn't he leave?

In interviews and press conferences, he explained again and again that he meant no disrespect to the military. He emphasized that he had friends and family in the police force. He stated that his concerns were not directed against the people serving the United States. Instead, his actions were a protest against what he saw as the nation's shortcomings in upholding its own values of equality and justice.

After speaking with military service members and other football players, he refined his stance. Instead of sitting, he would take a knee. It was a sign of respect, he said, for the men and women who had fought and died for the principles of his country: freedom, justice, and equality. He believed that taking a knee would help clarify his message and lead to more change.

The next game, one of his teammates, Eric Reid, joined him. Soon, dozens of others across the league did the same. The protest grew, spreading to other sports, from professional to amateur.

Kaepernick still played but he became known more for his activism off the field than his touchdown passes on it. And at the end of the season, he left the 49ers before they cut him.

In the years since, despite his being in excellent health, no other team has signed him. Many, including Kaepernick, believe that he has been blackballed for his stance.

Even so, he has no regrets about his decision to sit, then to kneel. He always knew that defiance had costs. But he also knew that for him the place and time was right, and he had to act.

What Kind of Situation Is This?

When deciding how to behave in a situation, especially if you experience tension, the first question of the Defiance Compass—*Who am I?*—helps us identify and connect with our core moral values. Once we do that, it is time to turn our gaze outward, to assess our surroundings and ask: *What kind of situation is this?* Like the first question, we might ask and answer it implicitly. But the more explicit we can be in our awareness and questioning, the more comfortable and practiced we can be with putting our defiance into motion.

The better we are at analyzing our surroundings, the more quickly and skillfully we can make informed decisions about defiance. We must assess for two main factors—safety and impact.

Is it safe for me to defy? (safety)
Will this action make any difference? (impact)

When it comes to defiance, we want to feel safe in several dimensions: physically, emotionally, financially, and socially. We also want to know that our defiance will be effective—that it will have a positive impact.

Think back to Rosa Parks, sitting in her seat on that Montgomery bus, looking once again into the indignant face of the bus driver James Blake. We know the journey that led to that day—the long struggle for justice, the late nights planning at the kitchen table, the black-and-white photos of the battered body of Emmett Till. Parks's actions were not the product of mere impulse. They were the product of close observation and clear-eyed assessment of her situation—not only within the struggle for civil rights, but right there on that bus, sitting among the other passengers, breathing their air.

Rosa Parks was well aware of what could happen if she refused to give up her seat. She knew the risks to her safety: She could be arrested, subjected to immense scrutiny, and face harassment and legal jeopardy.

But Parks also thought her refusal to move could have an impact. It could be a way to advocate for desegregation of the city's buses. As an activist with years of work behind her and impeccable credentials within the Civil Rights Movement, her fellow activists would rally around her. Her stand could make a difference.

Colin Kaepernick may have made a similar calculation. As a professional athlete, he had power, visibility, and a weekly televised platform. His motivation to sit during the national anthem may have initially begun as a small personal act of defiance, but as he quickly discovered, it also had impact. Although he didn't expect his image to go viral, it did—and soon, others joined him.

Although I recommend assessing for positive impact, defiance isn't always about winning others over. Sometimes, we want to defy simply to take a stand (or seat) for what we believe in.

The stages of defiance—tension, acknowledging discomfort, vocalizing it, then threatening to stop complying before the final act of defiance—can amount to a conversation with the environment. Even if you skip steps or go out of order, the stages help you test the waters and assess how defiance could affect you in the moment. These attempts to probe the situation help detect whether the time and place are right.

Colin Kaepernick kneeling under the bright stadium lights, Rosa Parks holding the number 7053 in her mug shot—these images are permanently burned into the retinas of many people around the world. Both of these people made a historic decision to defy on a particular day in a particular situation. The task we all face when we consider giving a True No is to be able to identify such a moment when it arrives—not through gut feeling or "instinct," but through careful, clear-eyed, and rapid assessment of the environment and one's place in it.

These are profoundly personal decisions. Some people defy despite great risks to their physical and psychological safety. Often, they defy even when they are certain it will not lead to lasting change. But a clear understanding of the possible risks, costs, and benefits allows us to make an informed decision. The choice is ours.

The Defiance Empathy Gap

Defiance is inherently risky. It will never be completely safe. A True No does not always bring with it the risk of bodily harm, but defiance can have immense psychological, social, financial, or professional costs. Going against the status quo could cost us our job or leave us isolated, mocked, or ostracized.

At the same time, a culture in which everyone is constantly swallowing their True No can have devastating consequences.

On an airplane, if a co-pilot cannot speak up about an error to the pilot in command, the lives of hundreds of passengers could be at stake, thousands of feet in the air. In law enforcement, if an officer has no way to report misconduct without fear of reprisals, the force and the public will suffer. In a hospital, a nurse who cannot flag a doctor's error without fear of being fired could lead to the death of a patient.

Several years ago, I interviewed fifty nurses and nurse managers at a major U.S. hospital for a research project. I asked the nurses questions such as: *How comfortable are you to speak up or raise work-related issues, especially those related to patient safety? What keeps you or other nurses in your unit from speaking up about problems?*

What I learned reinforced my concerns: Many nurses were not at all comfortable speaking up. The kinds of issues the nurses described as problematic were common but potentially serious: an error in medication, a chronic staffing shortage leading to shortcuts and mistakes.

By far the most common reason for why the nurses did not speak up was because of a lack of psychological safety. These nurses used the word *fear* often—they believed that disputing a nurse manager or a physician would lead to retaliation, and they were intimidated by the power imbalance inherent in their roles.

As one nurse put it:

> Probably one of the biggest things would be fear; fear of getting in trouble. . . . I think there is also a level of intimidation from higher-ups or . . . if there's someone at the bedside and the physician's saying "This is what we're going to do," and you know in your mind "This doesn't seem right" . . . That takes guts to step up and say, "Whoa doctor, no, we're not going to do that."

But fear was not the only limiting factor for these nurses. The other reason many of them gave was the belief that even if they did speak up, it wouldn't matter. The nurses didn't believe that speaking up would have any positive impact. As one nurse expressed:

> I think we all get to the point where it's like I'm not scared to speak up but nothing ever changes, nothing ever happens.

One can easily imagine keeping silent in such circumstances, even if you have witnessed an error. If you do not feel psychologically safe, *and* if speaking up wouldn't lead to real change, what is the point?

I asked the nurse managers similar questions: *How comfortable are the nurses in your unit speaking up or raising work-related issues, especially those related to patient safety? What do you think keeps the nurses from speaking up about problems?*

The nurse managers gave strikingly different reasons for why the nurses in their units did not speak up. They did not focus on psychological safety or impact. Instead, these higher-ups saw a nurse's failure to speak up as a personal failing—not a cultural issue. They believed that either the nurses abdicated their responsibilities for reporting errors, or they just did not have the ability—they were not skilled enough or confident enough to speak up. As one manager said:

> Some people, some nurses do not take as much ownership about what is going on in their environment with other patients.

In essence, the nurses thought that the hospital culture was the problem, while their bosses thought it was the nurses themselves who were at fault.

My colleagues and I named the discrepancy between nurses' and supervisors' perceptions the *voice empathy gap*. Managers and employees held different opinions about what prevented speaking up (voice), based on their positions in the organizational hierarchy. Because of their different roles, they interpreted the situation, and the people within it, in fundamentally different ways. Managers could not empathize with employees' emotions of fear and intimidation. They did not understand how uncomfortable it can be for someone to speak up.

Empathy gaps like these are problematic. If nurse managers like those I spoke with do not believe that there is a culture problem, they will do nothing to change the environment. And if nurses feel unsafe suggesting change, or believe that raising concerns will make no difference, a cultural stalemate can set in. And in a high-stakes setting like a hospital, everyone loses, particularly patients.

Empathy gaps are not limited to hospitals, and they are not always about speaking up. An empathy gap can apply to defiance more generally, resulting in a *defiance empathy gap*. Defiance empathy gaps are often at the heart of the question in Chapter Eight: Who Gets to Defy? What seems unsafe to you might be within a comfortable norm for those with more power—who then do not understand how you could feel unsafe. And when you consciously comply with a situation because you feel unsafe, those with power may misunderstand you, believing that you consented, or that you didn't take your responsibility to defy seriously, or that you simply lack the ability to enact a True No.

In other research conducted with my collaborator Catherine Shea, I examined the different perceptions between recruiters and candidates toward illegal demographic questions in job interviews. We identified a perception gap whereby interviewers often misconstrue illegal questions, about marriage and children, as benevolent gestures to build rapport or help the interviewee. In contrast, job candidates, especially women, perceive them as potentially discriminatory, leading to less trust in the interviewer and less desire for the job. This gap in perceptions persists because candidates, burdened with insinuation anxiety, tend to comply and answer the question. This response can inadvertently perpetuate the cycle where interviewers, oblivious to the negative repercussions of their line of questioning, continue with their potentially detrimental behavior.

Dismantling defiance empathy gaps is often a necessary part of creating an environment in which everyone's True No is valued. It is often difficult to do, but my research suggests that educating managers on the voice empathy gap can shift their beliefs. When managers were informed about the importance of psychological safety, situational constraints, and culture on speaking up, they were more able to understand their employees' perspectives.

For employees, however, creating change should not require you to constantly sacrifice your well-being in unsafe environments, even if those environments are exactly what you aim to change. You don't have to consistently fall on the grenade to make sure it doesn't hurt anyone else. When the time is not right, sometimes it is better to wait for a better opportunity—a time and place when your defiance will be less costly and can truly make a difference.

Knowing When the Time Is Right

In the beginning, Rachael Denhollander liked Dr. Nassar.

Pale, with short black hair and a weak chin, Nassar wore frameless glasses that he often pushed to the end of his nose, peering over their edges with an owlish expression. With his pleated pants, his tucked-in polo shirt, and his cellphone clipped to his belt, he looked more like the dorky dad of one of her gymnastics teammates than a world-famous doctor for Team USA.

At Denhollander's first visit to his office at Michigan State University, the doctor complimented her shoes and asked her about the science textbook she was carrying. He called her "kiddo" and said she could call him "Larry."

And when she told him about her pain, he actually listened.

He listened to her describe the pain in her wrists and her back, how it had become difficult for her to do even basic chores at home, much less the back handsprings her gymnastics coach asked her to perform at every practice.

Unlike other doctors she'd seen, Nassar appeared to know exactly what was wrong with her. He named the tendons that were inflamed in her arms and hands. He explained how Denhollander was hyperextending her wrists on every tumble, which in turn overtaxed the muscles in her lower back. He showed her how her hips were improperly rotated, and with the help of a model, he explained to Denhollander and her mother how he could manipulate the pelvic bone, pulling it back to where it needed to be.

Nassar had Denhollander lie down on her stomach, with her legs a foot apart. He placed one hand on her lower back, and the other around the inside of her leg.

He said he was going to apply some pressure. He did. But he also moved his hand under her shorts and underwear.

It was so brief Denhollander wondered at first if she'd imagined it. Then she doubted herself, thinking that this must be a normal treatment—she knew from discussions with her mother that there were internal therapies for pain. But she also knew that such procedures were not the first line of treatment for gymnasts, and that they were usually performed by certified specialists, and would require gloves, which Nassar wasn't wearing.

Denhollander stayed quiet.

She didn't say anything when Nassar did the same thing, just minutes later, during myofascial release. Or later again, during deep tissue massage, a technique involving pressure on specific trigger points.

In the moment, Denhollander froze. And later, she tried to reason away what was happening. After all, Nassar was so nice. He was one of the most famous doctors for gymnasts in the world. Surely he knew what he was doing. Surely he would never hurt her—and especially in front of her mother, who was sitting across the room, reading a magazine.

What Denhollander didn't realize in the moment was that Larry had positioned himself directly between her mother and the area he was touching, so her mother couldn't see what was happening.

What she didn't realize in the moment was that Nassar—for all his kindness, for all his acclaim, for all the awards and medals decorating the walls of his office—had been doing this for years, to hundreds of girls just like her.

And what she didn't realize in the moment was that he would continue to do it for many years to come, to many more girls.

Sixteen years later, Rachael Denhollander was a lawyer, a gymnastics coach, and a happily married mother of three. She had long given up her dreams of a career in gymnastics, but she adored her work, her children, and her hours at the gym, helping girls train in the sport she loved.

One morning, in the kitchen of her small Louisville, Kentucky, home while she was looking for her grocery list on her laptop with her teething baby daughter strapped to her chest, a headline leapt off the screen.

A BLIND EYE TO SEX ABUSE:
HOW USA GYMNASTICS FAILED TO REPORT CASES

Denhollander's world went silent. The article was in the *Indianapolis Star* and revealed how USA Gymnastics had been covering up sexual abuse for decades. Several gymnasts had come forward with allegations against a number of coaches and trainers, and Denhollander frantically scanned the list, looking for Larry Nassar's name. She didn't find it.

For all those years, she'd never gone public with what happened to her, multiple times, when she was fifteen. But that morning, she sent an email to the *Indianapolis Star*. She then went on to file a police report in East Lansing and a Title IX complaint at Michigan State, and in doing so, she became the first woman to publicly accuse Larry Nassar of sexual assault.

In the months to come, more than 150 women came forward with stories about how they'd suffered abuse at Nassar's hands. Among the group were some of gymnastics' most decorated athletes, Olympic medalists like McKayla Maroney, Aly Raisman, Gabby Douglas, and Simone Biles. For years, they had been silent. But now, emboldened by Denhollander's defiance, they shared their stories. And as a result, Larry Nassar will spend the rest of his life in prison.

So what changed for Rachael Denhollander? How did she know it was time to act?

For one thing, she had had time to process the abuse she had undergone. She was no longer confused or uncertain about what had happened to her. With time had come perspective and clarity. And, as a mother herself, she could not bear the thought of someone like Larry Nassar hurting children like hers—or gymnasts like the ones she coached.

Denhollander was also older and knew the setbacks she might face. When she was eighteen, a few years after Nassar had abused her, she told a nurse about it. She also confided in her coach. But nothing happened.

"No one else is saying what you're saying," her coach told her.

Back then, it had been her word against his. And Denhollander's word clearly didn't have much weight. Who would believe a teenage girl accusing a beloved doctor of something so heinous?

Now, though, as an adult and a successful attorney, she had more power, and she knew how to build a case. As she wrote in a *New York Times* op-ed:

I came as prepared as possible: I brought medical journals showing what real pelvic floor techniques look like; my medical records, which showed that Larry had never mentioned that he used such techniques even though he had penetrated me; the names of three pelvic floor experts ready to testify to police that Larry's treatment was not medical; other records from a nurse practitioner documenting my disclosure of abuse in 2004; my journals from that time; and a letter from a neighboring district attorney vouching for my character. I worried that any less meant I would not be believed—a concern I later learned was merited.

It had not been safe for her to speak up as a teenager. And it was not totally safe for her to speak up as an adult. After going public with her accusations, Denhollander at first endured a harrowing gauntlet of public scrutiny. Her church abandoned her. Many of her friends deserted her. She struggled with the loss of her privacy, and with the facts of the abuse she suffered spelled out in excruciating detail in the press.

But her family and some of her closest friends provided her support. And she felt strong enough to withstand any pushback she received— because she knew that now, her words could make a difference. She had assessed her environment, and she believed she now had sufficient safety and potential for positive impact. And she sensed that there were enough other women who had experiences similar to hers—and who would be willing to join her.

Denhollander was different now and so were the other women and their environments. These assessments featured in Denhollander's defiance calculus to allow her to identify the right time for her defiance to be both safer *and* have impact.

She didn't believe her words would make a difference at age fifteen.

But sixteen years later, she knew that they would.

The time to defy does not need to be perfectly right; it just needs to be right *enough,* safe *enough,* or effective *enough.*

Otherwise, we can always find an "out," excusing our inaction in the name of safety:

Things could get so much worse for me.

This is the price we pay to survive here.

Or impact:

There's nothing I can do that will make any difference, so I may as well comply.

Nothing I say or do will ever change things.

These rationalizations may seem to arise from a clear-eyed assessment of safety and impact, but if they are never interrogated, they can grow into a pessimistic denial of our agency—a perennial rationalization for nonaction. They can turn potential resistors into defeated conformers.

We assess our environment for impact and safety not to forestall defiance, or to postpone it indefinitely with a self-defeating shrug, but to ensure its power and effectiveness. When we test the limits of a situation, analyzing our safety and power to make change, we are not asking *if* we should defy but *when*—not *whether* but *where* and *how*.

Transforming Cultures

Denhollander acted years after leaving competitive gymnastics as an athlete. But her voice has already had a broad and lasting impact on how people see the environment and institution that allowed Nassar's abuse to go on unpunished for decades. Her goal, like Michael Quinn's—the retired police officer—was to reform a broken system. But her voice was only effective after she left the situation.

What about people like Kevin, the current police officer, who are still living and working within those broken systems? What are the challenges and possibilities of their True No?

"The people make the place," psychologist Benjamin Schneider wrote in his famous paper explaining how organizations naturally end up with similar people exhibiting similar personalities. He called this the attraction-selection-attrition (ASA) cycle, in which people are *attracted* to a particular type of organization and attempt to join; they are *selected* to become members if others in that environment believe they "fit"; once hired, if they don't fit, they leave, either by their own will or the volition of others—and their *attrition* further homogenizes the initial environment.

This model highlights how difficult it is to bring about cultural change in an organization. If everyone who doesn't fit leaves, then the culture will remain the same.

But defiance can give us a tool to change cultures, not simply leave them. Remember, we have several options *besides* a principled exit when our values do not align with an environment. Defiance doesn't have to be loud to make change, and leaving isn't the only way to defy. Through routine advocacy, dissent, and professional disobedience, those who don't currently "fit" can help to change a culture for the better, working from the inside. Michael Quinn, at great risk, defied his fellow police officers when he saw their corrupt practices. Breaching the blue wall, his defiance changed the behavior of the officers around him, leading to a cultural and ethical shift in his environment.

Cultures are never totally static. The people may make the place, but that doesn't necessarily mean by leaving, but by *staying* and *defying*. Defiance can change even the most hidebound and regimented environment into something new. All of us have this power and using it does not necessarily mean we must commit ourselves to grand gestures. Defiance does not need to be "big" to be impactful. Small defiant acts matter more than you think.

As organizational psychologist Karl E. Weick wrote, small wins—"a series of concrete, complete outcomes of moderate importance"—can, over time, add up to major victories. Even the earliest stages of defiance can lead to major change.

Perhaps, by providing some extra information, a nurse can plant a seed of doubt in a physician's mind about whether a diagnosis, based on too brief an examination, is an accurate appraisal of a patient's condition. By requesting clarification about an action, a police officer might prompt a superior to reexamine his behavior. A simple "What do you mean by that?" after a sexist or racist remark can encourage a co-worker to focus on their words and hopefully discourage them from repeating them in the future.

Curiosity can be an effective way to approach others, opening pathways to change. Asking questions can decrease tension and reduce defensiveness, encouraging a collaborative exploration of a topic. When people know that you are listening to them—not judging them or adopting a position of moral superiority or righteousness—they are less likely to become entrenched in their position. They might even change their minds or behavior.

These moments might seem small, but they are the building blocks of cultural and societal change. After you have experienced stages one and two of defiance, feeling tension and acknowledging it to yourself, you have a decision to make: keep your tension to yourself, or acknowledge it to others. If you proceed to stage three of defiance—escalation and vocalization—you have an opportunity to test the environment and even begin to change it.

Asking for clarification raises the volume on a situation, and when you raise the volume, you can begin to transform it.

You may find that the situation is not as hopeless as it seems—that it is more malleable, that your small act of defiance carries more weight than you thought. You may also find that you are not alone, and perhaps others are willing to speak up with you. Either way, you learn valuable information that can help you decide whether it is time to defy. It might never be exactly the "right" time and place—but that doesn't mean it's necessarily the wrong time and place, either.

What Dal and Defiance Have in Common

My parents are sophisticated connoisseurs—and cooks—of delicious Indian food. So when they order takeaway, their standards are quite exacting. When ordering from a new restaurant, they have devised a simple and efficient test to see if the food will meet their threshold for quality. On their first order, they always get the same thing: a basic lentil dal.

It's a simple dish, one that almost any decent Indian cook could make in their sleep. That's why my parents use it as a barometer of quality. If a restaurant can't get a basic dal right, there is no way that they'll succeed in the more complex dishes of flavorsome Indian cuisine.

Choosing a restaurant for delicious takeaway does not typically require defiance. But my parents' test is a terrific example of a process necessary for evaluating quality: active and rapid assessment.

So much of the way we respond to the world escapes our conscious attention. Moment to moment, second to second, we are bombarded by sensory information. From the air temperature to the color of the walls to minute fluctuations in someone's vocal tone, thousands of sources of information continuously affect how each of us interprets our environment. Advertisers, architects, and social engineers alike know this. For example, some restaurant owners attempt to steer customers toward certain orders with carefully crafted menu placements.

In an environment that might require defiance, without active assessment, we are much like those unwitting diners, blithely consuming whatever stands out to us. But when we take the time to consciously evaluate where we are and who we are dealing with, we can learn to more quickly determine whether it is safe to defy, whether it could have impact, and who—if anyone—to trust.

With repeated practice, we can become more skilled in analyzing our environments. With experience, like my parents assessing food from a restaurant, we can develop shortcuts to evaluating a situation, based on specific context cues, which allow us to act more rapidly and accurately, and to answer the questions—*Is it safe for me to defy? Will this action make any difference?*

Your decision might be different in a boardroom with your boss than at your kitchen table with your friends. But it is fundamentally based on the same rapid analysis—one that will get quicker with each repeated exposure.

We may, over time, develop an intuitive expertise in defiance: to be sure, this is not a "gut feeling" but rather a sped-up assessment, borne of much experience and practice in certain contexts. We might know, from frequent observation, that our boss has a quick temper in certain situations—and that defiance, even if righteous, might be met with angry retaliation. Or we become aware that our friend has a "tell" whenever he is about to share an inappropriate joke—and that moment is the perfect time for pushing back.

Like my parents assessing their dal, repetition and observation lead to knowledge. We can *prepare* for defiance by analyzing situations and anticipating how we might react when we are asked or told to do something that goes against our values. And the more we *practice* speaking up, even in small ways, the better we get at assessing how our defiance affects—and is affected by—our environment. Small-scale defiance develops our skills, increases our confidence, and prepares us to recognize the right time and place.

Just as there is no perfect defiant person, no foolproof revolutionary who has it all figured out, there is no perfect environment for defiance—no failsafe formula of risk, impact, and safety that can tell us exactly when and where to defy. But we can do more to ensure that we see a situation for what it is. Through preparation and practice, we can progress: the repeated exercise of our defiance muscles helps us better recognize when it's time to defy, better understand our environment, and better appreciate when a True No is no longer just our preference but our responsibility.

12

The Superpower of Responsibility

The morning of May 24, 2022, started out as a proud one for Angeli Rose Gomez. After a successful academic year, her two sons were graduating the second and third grade at their elementary school. There would be a short ceremony with a photo opportunity. At first Gomez told her boys she would have to miss it—as a farm supervisor, the only way she could make it in time would be to come straight from the fields, and she didn't want to show up dusty and dirty at such a momentous occasion.

But her sons wouldn't take no for an answer.

"Mom, we want to take this photo," they insisted.

So she changed her mind. On graduation day, she put in several hours of work early in the morning, then left the fields and drove forty miles to Uvalde, Texas, arriving just in time to see her sons receive their certificates. She wrapped her arms around her boys and beamed proudly as a photographer took their picture. Then she drove straight back to the farm to continue her day of work.

Ten minutes after arriving back at her job, Gomez received a panicked phone call from her mother telling her there was a shooting at the school.

Within seconds, Gomez was back in her car, careening down the highway. Instantly she forgot about her dusty clothes, her unfinished tasks at the farm, the confused looks on the faces of her co-workers watching her leave for the second time that morning. In her terror, all other thoughts were eclipsed by her need to get back to Uvalde as fast as possible. She sped past other vehicles on the road, her mind narrowed to a single objective: getting back to her sons' school. The parched brown fields flew by her windows, other cars and pickup trucks vanishing almost as soon as they appeared. In record-breaking time, she was once again entering Uvalde. With a tense grip on her steering wheel, she prayed that it would be fast enough.

Gomez pulled to a stop near the front entrance of Robb Elementary School. Within seconds, her car was surrounded by U.S. marshals in bulletproof vests and tactical gear, their guns at the ready.

"Ma'am, you'll have to move your vehicle," one of them told her. "You're not allowed to park here."

Farther back, she could see a crowd of obviously distraught parents screaming at police officers, pleading with them to do something, *anything*, to save their kids.

Gomez looked at the marshals, then the parents, then the line of Uvalde police officers standing at the fence. No one seemed to be entering the school, even though it had been half an hour since she'd gotten the call about a shooter. Everyone was just standing around.

"Is anyone going in?" she asked. "My kids are in there."

The marshals didn't answer, just repeated their insistence that Gomez move out of the way or they would have to arrest her for being uncooperative.

Gomez could hear the parents' frantic cries of anxiety and shouts of desperation at the marshals and police who held them back from entering the school.

"They need to go in there," one father was yelling. A mother shouted out, "Shoot him or something!"

When a parent rushed the fence, attempting to push it over, an officer tackled him to the ground. When another father tried the same thing, he was pepper-sprayed.

The crowd was inconsolable now, the cacophony of distressed screams and yells from frightened and frustrated parents filling Gomez's ears. She added her voice to their anguished chorus, berating the officers for not pursuing the gunman. What were they waiting for? Why weren't they doing their jobs?

Gomez had been terrified, then confused. But now she was furious. Where were her children? Were they alive? And why was no one doing anything to save them?

"Ma'am," a marshal told her sternly. "If you don't move, we'll have to arrest you."

"Well, you're going to have to arrest me because I'm going in there, and I'm telling you right now I don't see none of y'all in there," Gomez yelled. "If y'all don't go in there, I'm going in there."

This was not the response the marshals were looking for. They immediately pinned her arms behind her back and slammed handcuffs on her wrists. The horrific irony was not lost on Gomez: She, an unarmed, petite mother, was being arrested at a school shooting, forced into submission by armed police officers nearly twice her size.

And all this time, she could hear gunshots echoing across the parking lot, and the agonized screams of other parents being held back from the school by the police.

Gomez pleaded with the officers to move faster, to enter the school, to save her sons. She imagined them cowering in their classroom, frightened and asking for her. Her oldest son, his watchful eyes wide in terror, staring at the door; her youngest, hiding his face behind his shoulder-length curls. Straining against the handcuffs, bruising her wrists on the metal and twisting her arms in desperation, she begged and pleaded, cajoled and harangued. But nothing she said made any difference, and eventually, she realized that continuing to resist in this way wouldn't get her anywhere.

So she took a different tack. Recognizing a local Uvalde police officer, she took several deep breaths and consciously lowered her voice.

Calmly, she persuaded the marshals that she would cooperate and that they could uncuff her.

The local cop stared at her for a moment, then nodded.

"You can unlock her," he said. "She's calmed down."

Gomez was compliant and quiet as the marshal moved behind her with the keys. She relaxed her tense body as he fit the key into the lock. But in the split second after her hands were freed, she wriggled free of the marshal's grasp. And instead of walking back to the other parents, she sprinted directly toward the school.

Gomez had a few seconds' head start. She knew that the marshals and Uvalde police officers were pursuing her. But unencumbered by weaponry or tactical gear, she was faster and more agile than they were. She swiftly jumped the metal chain-link fence that lay between her and the classrooms where her terrified children were hiding. Her heart racing, she dashed to her older son's classroom door and slammed her fists on it until the teacher appeared.

"The police are behind me," Gomez said. "They're already bolt cutting the fence to get me."

"You think we have time to get out?"

Gomez responded, "You have time!"

The teacher opened the external door leading out of the classroom and cautiously beckoned to her students. Gomez didn't have time to do what she wanted to do: run to her oldest son, sweep him up in her arms, and tell him everything would be okay. The police were almost on her, and if she didn't keep moving, she'd never get to her other child.

"I'm going to run for my other son," she said.

Then Gomez kept running, as fast as she could. She could still hear the gunfire in the distance and kids screaming. As she passed nearby the cafeteria, she heard cries, "I want my mom! I want my mom!"

She fell to one knee and said a quick prayer. Then she ran again.

Through each window she passed, she saw the eerie vision of a school in lockdown. She didn't know exactly where her younger son's classroom was. Evading the pursuit of the marshals and police officers, she dashed around the corner of the school until finally she found it.

She banged on the door and yelled for everyone to run. But the teacher wouldn't let her in. Gomez screamed her son's name, jiggled the handle, and threw her weight against the door, trying to break it down with her small frame.

But it was no use, and within seconds the police caught up with her. She heard the sound of their heavy boots behind her, felt the rush of air as officers in full body armor came to a stop behind her.

She could no longer bang on the door. Her arms were restrained, as they attempted to escort Gomez out of the school. A voice yelled in her ear that she was putting her life in extreme danger.

"Give me a [bullet-proof] vest then," she yelled. "I'm not leaving until you evacuate the school."

Seeing the officers, the teacher unlocked the classroom door, and a stream of children suddenly flooded out.

"Mom!" Gomez's younger son cried, his arms open wide to her, his face streaked with tears beneath his curly hair. "Mom!"

Nothing could stop her now. Gomez again slipped away from law enforcement and ran to her son, grabbing his hand. Together, they bolted for the relative safety of the parking lot, where her other son was waiting.

Her chest heaving, sweat pouring from her body, Gomez was too exhausted and keyed up on adrenaline to cry. But when the two brothers saw each other, they burst into tears. They hugged each other, then their mother.

"I'm so glad you're okay," the older boy said, reaching out to hug his younger brother.

He had always looked so much like his mother—the almond eyes, the pert nose, the expressive mouth—and at that moment their expressions were identical, a mix of terror, concern, and relief.

"I was so worried that you weren't," her younger son replied.

The events of May 22 at Robb Elementary qualified as the third most deadly school shooting in American history. Two teachers and nineteen students were killed.

Although other people witnessed her actions on that day, Angeli Rose Gomez did not tell her story at first. A local Uvalde police officer quietly informed her that because she was still on probation for an offense committed nearly a decade prior, speaking to the media could lead to new charges related to obstruction of justice. It took a judge reducing the terms of her probation and commending her actions on the day of the shooting to give Gomez the reassurance she needed to publicly tell her story.

Gomez is still angry about the way she was treated on the day of the shooting. But more importantly, she is saddened and traumatized by how unresponsive the police were to the tragedy unfolding in front of them. In the days and weeks after the shooting, she joined those protesting the police force in downtown Uvalde, a small town west of San Antonio that does not usually see much in the way of civil resistance.

Every day, she wonders what would have happened to her boys if she had not defied police orders. Would they have survived? Would she be able to live with herself if they hadn't?

One thing she knows for certain: The police were negligent. They stood by and did nothing while children died. As she tearfully told a reporter, "They could have saved many more lives, they could have gone into that classroom. . . . They were more interested in keeping us out than getting into that school."

Gomez said that in the months following the attack, she was harassed by local police officers to such an extent that at one point she had to live separately from her children, to spare them the threats and surveillance she experienced. Several times, she has been followed by police. Once, through the windows of her home, a squad car began flickering its lights at her—a clear sign, she says, that local police were targeting her for intimidation.

But she keeps pushing, keeps telling her story, because she wants people to know how badly the police failed the children of Uvalde. She has given interviews to the national press, marched in protests that successfully resulted in the firing of the Uvalde school police chief, and has threatened a lawsuit against the Uvalde Police Department, claiming harassment. Angeli Rose Gomez wants to make sure that the people in charge that day are never in positions of authority again. She believes she owes it to her sons to protest in this way—she owes it not only to them, but also to the parents of all the other children of the school, especially those who did not make it out alive that day.

What Does a Person Like Me Do in a Situation Like This?

Angeli Rose Gomez's actions might seem superhumanly courageous. But she was not the only parent to attempt to rush Robb Elementary. Others were pepper-sprayed, tackled, and tased.

"They didn't do that to the shooter," Gomez told *The Wall Street Journal*. "But they did that to us."

There is a reason why these parents were willing to charge a police barricade, ready to run into a school where an armed gunman was firing rounds from an assault rifle. When our loved ones face imminent harm, many of the usual barriers to defiance that can stop us in our tracks seem to crumble. Many of us find it easier to defy authority if we are doing so for someone else: a loved one, a partner, a child. We are less likely to succumb to the internalized pressures for compliance that guide so many other areas of our lives. We feel less insinuation anxiety. We are, in short, willing to fight for others in a way that we are often less willing to fight for ourselves.

As a parent, I find this to be true. It is so much easier for me to overcome my resistance to resistance and defy when the well-being of my son is at stake. I often find myself acting for his benefit in a way that I'm unable to do for myself. In contrast to the struggle I felt to reject the CT scan my doctor recommended, for example, I can more easily reject unnecessary medical procedures for my son and seek multiple opinions for his welfare when I need to. Even though I spent most of my childhood steadfastly obedient toward my teachers and school rules, I experience relatively less distress raising concerns about my son's school's policies if I believe they are detrimental to the well-being of the children. These actions sometimes don't even register to me as defiant.

But they are defiant. And they show that so many of us can act defiantly when we feel responsible for someone else. We don't even think about it. Like Angeli Rose Gomez, when it's for our children, we run for the fence.

When she heard there was a shooter at her sons' school, Angeli Rose Gomez hurdled over every roadblock to resistance. Unlike the police officers at the scene, she was not under a hierarchical chain of command. As an outsider, she could see the situation in stark clarity: An armed murderer on a rampage was firing shots inside a school, and the police were standing outside, refusing to engage. This was unacceptable to her, a horrifying status quo she had no interest in upholding.

For Gomez, there was no hesitation and no second-guessing her decision. The solution was obvious to her: She would charge into the school, despite police orders and the risks to her own life, because it was her *responsibility* as a parent to protect her children. She quickly assessed her safety, continuously registering how far away the gunshots were from her, and she ran to her children.

Who Am I?

What Kind of Situation Is This?

What Does a Person Like Me Do in a Situation Like This?

Angeli Rose Gomez knew the answers to the Defiance Compass questions almost instantaneously. She knew who she was: a mother who values her children's lives above her own. She knew the situation: It was dangerous, but her children were already in danger and she might be able to help them. And she knew what she had to do in such a situation: She had to act immediately to save them, even if it meant openly defying law enforcement, even if it meant losing her own life. It was a decision that reverberated with parents not just in Uvalde but around the country.

Often when we fail to defy, we let external and internal pressures cow us into submission. We may doubt our ability to take action or displace our responsibility onto others.

But sometimes, situations that call for defiance require us to think outside ourselves, to imagine our actions on a wider scale. A True No does not only affect us; the decision to defy can have ripple effects for many others, regardless of whether it is a dangerous, life-threatening situation—as it was for Angeli Rose Gomez—or a seemingly harmless, low-risk one—like refusing to stand during the national anthem. Angeli Rose Gomez shows us the immense galvanizing force of responsibility, its power to spur us into action. She took responsibility both at the time of the shooting and afterward with her activism. Her story shows that defiance is an individual act with communal effects.

Embrace Responsibility

When people in Stanley Milgram's obedience experiment were asked why they continued to administer shocks, one of the most common responses was simple—they were just doing what they had been told. Milgram saw abdication of responsibility as one of the most pernicious, and potentially dangerous, effects of authority on ordinary people. As he put it:

> The most common adjustment of thought in the obedient subject is for him to see himself as not responsible for his own actions. He divests himself of responsibility by attributing all initiative to the experimenter, a legitimate authority. He sees himself not as a person acting in a morally accountable way but as the agent of external authority.

The "just following orders" defense, Milgram noted, was also the preferred response from the Nazis who were tried for war crimes at the Nuremberg trials. But that does not necessarily mean that it is a lie, a false statement meant to exculpate someone from guilt.

"Rather," he wrote:

> . . . it is a fundamental mode of thinking for a great many people once they are locked into a subordinate position in a structure of

authority. The disappearance of a sense of responsibility is the most far-reaching consequence of submission to authority.

But just because someone obeys authority does not mean they've lost their sense of morality. They have simply redirected it toward an authority figure. Instead of considering whether or not their own actions are moral, an obedient person measures his morality in relation to how well he has obeyed.

Employees, for example, can become so narrowly focused on doing a good job that moral concerns can fade into the background. When we aim only to please our boss, our frame of responsibility shrinks to the walls of our cubicle and the confines of our company. We push any ethical dilemma to the background or even remove it from our decision-making altogether—a phenomenon psychologists call *ethical fading*. We lose ethical perspective, jettisoning all connection with our higher principles and deepest values, because we have located our sense of responsibility not to ourselves or others, but solely to the immediate authority.

In Milgram's experiments, some of his subjects exhibited no flicker of a moral quandary in administering shocks. One, a thirty-seven-year-old welder named Bruno Batta, was a participant in one of the experiment's proximity variations. Instead of sitting in a separate room, Batta was seated next to the learner, and after the 150-volt level, he had to physically place the restrained man's hand onto the plate administering the "shock." This he did with almost complete indifference to the learner's cries of pain and anguish, assuming what Milgram described as a "rigid mask," his forceful motions displaying "robotic impassivity."

Over and over, Batta forced the learner's hand onto the shock plate, ignoring the learner's pain while speaking politely and deferentially to the experimenter. Only once did he address the learner, and that was to chastise him for refusing to answer.

After the experiment was over, when asked who bore responsibility for the learner's pain, Batta at first did not understand the question. In the end, he said responsibility lay with the experimenter.

"I had to follow orders," he said.

Milgram named the phenomenon exhibited by Bruno Batta "the agentic state"—it describes a person who resigns themselves to being the unthinking agent of another. In such a state, a person no longer views themselves as responsible for their own actions but defines themselves as a vehicle for carrying out the wishes of others.

Defiant people, by and large, don't abdicate responsibility for their actions. Indeed, many of the people who defied in the Milgram experiment did so because they felt responsible for the suffering they believed they were causing.

After hearing the victim verbally complain at 150 volts, thirty-two-year-old Jan Rensaleer, an industrial engineer from Holland, first questioned the experimenter: "What do I do now?" Jan was distressed when told to continue, holding his hand to his head and glancing repeatedly at the experimenter during the following questions.

But at 255 volts, he pushed back from his station, turned to the experimenter, and adamantly refused to administer any more shocks. As an engineer, he knew the effects such shocks could have on the body and the mind. He understood the kind of suffering he was administering.

Jan was deeply troubled by what he'd done, but he made no attempt to displace responsibility for it back to the learner or experimenter. Instead, he took all of the blame himself.

"I would put it on myself entirely," he said, when asked who was responsible for the learner's pain. "I should have stopped the first time he complained. . . . One of the things I think is very cowardly is to try to shove the responsibility onto someone else."

Jan had lived in Nazi-occupied Holland during World War II, so he had a special connection to the mission of Milgram's experiments. So did thirty-one-year-old Gretchen Brandt, a medical technician from Germany. She refused to shock the learner after the 210-volt level, explaining that she "did not want to be responsible for any harm to him."

Jan was so moved by his experience in the lab that he later wrote to Milgram, offering his services. He knew what happened when people were allowed to abdicate responsibility for immoral actions; he had personal experience of what could happen when people followed orders without question above all else. And he felt a responsibility to ensure that such an atrocity as the Holocaust never happened again.

"Although I am . . . employed in engineering," he wrote, "I have become convinced that the social sciences and especially psychology are more important in today's world."

Duty to Defy

Milgram's concept of the agentic state emerged as a particularly seductive idea, one that could explain everything from Nazis to company "yes-men." But the concept tends to discount the numerous signs of distress and attempts at defiance exhibited by many of Milgram's "obedient" subjects.

Understanding defiance as a spectrum allows us to honor the in-between moments—the nervous laughter, the sad sideways glances at the experimenter, the pauses, the questioning that stops short of actual threats of refusal. In these tentative stages of defiance, we can see glimpses of people trying to defy. When we fail to defy, it's often not because we don't know our values but because we don't know how to enact them and proceed through the stages of defiance.

Those who do defy are able to do so for a number of reasons. They not only feel safe enough to defy; they connect with their values and recognize their free will. Ultimately, they feel *responsible* for their own actions, and they believe in their *ability*—that they have the necessary skills and confidence to defy.

This feeling of responsibility does not dissipate simply because someone else gives you orders. And although the legal system holds those who make the decision more culpable than those who were "just following orders," the reality is not so simple.

In several studies I conducted with my colleague, Kaitlin Woolley, we found that feelings of responsibility actually *increase* when a person is swayed by another to make a decision that goes against their better judgment and a negative outcome occurs. We call this the *kicking yourself syndrome*—people feel more culpable when they think they should have known better than to follow bad advice. If those in the Milgram experiment had been informed they had killed the man in the next room, you can imagine their regret:

I knew this was wrong.

I knew I shouldn't have done that.

Why didn't I defy?

We found similar remorse from people who decided to go along with someone else's bad recommendations, asking themselves, "What if?" These are the kinds of thoughts that haunt people after they take poor-quality, conflicted, or unethical advice: *What if I had made another decision? If only I had said no.*

We might follow orders because we think the responsibility and blame will lie with another person if something goes wrong. But in fact, following someone else's advice against our better judgment doesn't save us from feeling culpable. Quite the opposite; we actually feel worse. We cannot wish away our responsibility, or palm off blame to another. Better then to stop running from responsibility and be proactive about using it to align our behavior with our values.

Zusha!

Defiance can be a world-changing decision. When we take our responsibility for others seriously and enact a True No, we have the power to remake corrupt institutions, reform oppressive environments, and even topple authoritarian regimes. The onus for societal change, however, should not lie at the doorstep of individuals who are at the receiving end of oppression or discrimination. Those within institutions, especially those with power, have greater responsibility to act in ways that preserve a person's right to a True Yes and a True No.

The gymnasts who spoke up about Larry Nassar's abuse were the victims not only of one doctor who robbed them of their consent, but of a system that devalued their agency and silenced their defiance. Taught to ignore their pain, to overlook their injuries in pursuit of victory, to sacrifice their bodies for the sake of the team, they had been stifled and controlled by an organization that took their compliance for granted.

That was one of the factors that made Rachael Denhollander's—and then so many other gymnasts'—defiance so difficult. To be defiant, to speak up about what had happened to them, required that they not only navigate a conflict between their interdependent and independent selves, but also that they oppose an entire institutional structure much larger than any individual, and one to which they had dedicated a significant portion of their lives. Speaking up about abuse in Larry Nassar's exam room was more than just exposing a single abuser—it meant shining a light on an entire abusive culture.

After the gymnasts told their stories, Larry Nassar was criminally charged and sentenced to spend the rest of his life in prison. Béla and Márta Károlyi, the much lauded coaches who were accused of turning a blind eye to Nassar's abuse, were forced to close their training ranch in Texas, where some of the assaults took place—and where, for decades, gymnasts had alleged harsh, borderline abusive training sessions. And USA Gymnastics instituted a number of reforms, including drafting an Athlete's Bill of Rights, ensuring more comprehensive screening of coaches, and increased training to prevent sexual abuse. Whether these reforms are enough to ensure future gymnasts feel safe, sufficiently protected, and free to give their True Yes and True No is yet to be seen.

Creating successful systemic change requires those responsible to change their behaviors. One success story can be found on the public buses of Kenya. Each year worldwide, more than one million people are killed and between twenty and fifty million are injured in motor bus crashes, many of them in the developing world. The problem is especially acute in Kenya, which has a well-developed public transportation network of *matatus,* or minibuses. The overwhelming cause of matatu collisions is speeding, compounded with the fact that most passengers do not feel comfortable confronting drivers about their reckless driving. Passengers may feel some insinuation anxiety—the concern to imply that their driver is not doing a good job. Or they may feel the driver would ignore them or drive even more dangerously. Whatever the reason, passengers are often not willing to ask their drivers to slow down or drive more safely.

To be sure, the responsibility to drive safely is on the drivers and those responsible for hiring the drivers. But making room for honest feedback can spur better behavior. To make passengers feel comfortable speaking up to their driver, a pair of researchers, James Habyarimana and William Jack, came up with a simple idea: stickers placed prominently in buses, where any passenger could see them.

ZUSHA! each sticker reads. In Kiswahili, this means *PROTEST!* or *SPEAK UP!*

You have the power to slow down a reckless driver, the sticker continues.

In collaboration with a major Kenyan insurance company, these researchers placed the stickers in approximately eight thousand buses. They soon noticed striking results: average vehicle speeds were down, and so were crashes compared to the control group with no stickers. Between 2011 and 2013, there were an average of 140 fewer wrecks per year in vehicles with the stickers. That represented a reduction of 25 percent, or fifty-five lives saved per year.

Perhaps passengers felt more empowered and responsible to speak up, and when they did, perhaps the drivers were more likely to listen to them and do as requested. Or maybe the mere presence of the placards made the drivers *anticipate* that their passengers might object to their reckless driving, thus changing the way they operated the vehicle. It is difficult to say, but the results speak for themselves: By displaying stickers and informing passengers that they had the power to speak up and enact change, social norms changed, which resulted in a safer environment for everyone.

Efforts like this show the power of encouraging defiance in everyday life. When people are given space to enact a True No in psychologically safe environments, they can help create a world in which others are safer and more respected.

Individuals and institutions can create substantial change not through uncoordinated siloed attempts, but by working together. Whether it is USA Gymnastics, police departments, or a public bus line, institutions have a responsibility to both the people that comprise them and to the people they serve. The onus for systemic change must be on institutions to create cultures that allow for defiance and respect individual autonomy. But individuals, acting in accordance with their values, connecting with the responsibility to themselves and to others to defy, also have power. Sometimes our defiance may feel like a lonely independent act, but it is in fact a fundamentally *interdependent* act, one that brings us into community with the wider world.

13

Level Up

About a year after my unnecessary CT scan, I began to feel a growing discomfort in my shoulder following a routine vaccination. As the weeks went by, the pain worsened, making it frustratingly difficult for me to perform everyday tasks like getting dressed. So, to get a referral for physical therapy, I booked myself an appointment at an outpatient orthopedic clinic, near Georgetown University in Washington, D.C., where I was teaching at the time.

After I had filled out a small forest's worth of paperwork in the waiting room, my name was unceremoniously called by a young medical assistant dressed in pink scrubs and colorful sneakers. She led me down a hallway, and it was only when I saw the word "X-Ray" on a sign above the door that I realized she was not, in fact, taking me to the doctor's exam room.

"Where are we going?" I asked.

"To get your X-rays," she answered.

"But I haven't seen the doctor yet," I said, puzzled.

The assistant realized I had stopped walking. She turned to look at me, her eyes questioning my lack of compliance.

"All new patients have an X-ray before seeing the doctor."

After a few seconds, which felt like an eternity, I repeated what I'd just said to her in a level voice: "I haven't seen the doctor yet."

She cocked her head, seemingly unable to comprehend what I had said.

"Are you refusing an X-ray?" she then asked incredulously.

This seemed to be an accusation. But my training as a physician was clear: First see a patient, take a detailed history, perform an examination, and consider the differential diagnoses—the list of possible conditions explaining the patient's symptoms. Only then, consider investigations (such as X-rays) required to get closer to a final diagnosis, and only if it would change the treatment plan.

I knew there was no medical reason to take an X-ray first, and I wondered why *all* new patients would require an X-ray. I didn't like this feeling of being on a conveyor belt, shuttled from room to room, powerless and out of control.

"Yes," I said.

We stared at each other. Furrowing her brow as she glanced toward the X-ray room and then back at me, the assistant seemed, for a moment, more uncomfortable than I was. She clearly had no idea what to do with me.

Finally, she led me back to the waiting room. "Wait here," she instructed.

I obeyed. Other patients came and went. I stayed in my seat, shifting uncomfortably on the worn green vinyl. Although I tried to read one of the months-old magazines on the racks between the chairs, I couldn't concentrate. Anxious thoughts started to run through my head:

I'm not fitting in with how things are done.

I'm making a scene.

I'm being a "bad" patient.

But I also remembered that unnecessary CT scan that, against my better judgment, I'd complied with the previous year. How powerless I'd felt. I had resolved, after that experience, not to comply with unnecessary procedures just because I felt pressure. I'd imagined how I would act differently in the future, in a similar situation—a situation like this one.

I was finally called to see the doctor, a middle-aged, slim white man wearing a thin tie and a disgruntled expression. His first question was dripping with barely disguised hostility.

"No X-ray?"

"No," I answered. "I want you to examine me first. I think I have chronic inflammation."

The doctor did not appear to want to conduct an examination. It was almost as if he felt repulsed to touch me, his patient. But he reluctantly obliged, awkwardly prodding my shoulder before hurrying back to his stool after mere seconds. The exam was so cursory, he didn't even examine my range of motion.

"You need an X-ray," he repeated, gazing down at my chart.

"Why?" I asked. "Inflammation wouldn't show up on an X-ray."

"Well . . . to make sure you don't have anything . . . *boney* going on," he said, after a short pause.

"Boney? Such as what? What's your differential diagnosis?"

He paused again and answered, "Bone cancer."

My eyes widened.

"Bone cancer? You think I have *bone cancer*?"

"Oh no, no . . . ," he said, stuttering. "This . . . is just how we practice medicine here."

An uncomfortable silence ensued. This man knew I had trained as a doctor in the U.K., and we both knew that this was an unacceptable— even reprehensible—answer.

The doctor didn't seem to know what to do with his hands. He ran them through his bristly hair; he fidgeted with a pen. He gazed down at my chart again, as though he would find something there to bolster his position, then knocked on it with his closed fist.

"You need the X-ray," he said, raising his voice. "All patients here get one. It's just how we do things. And now, since you've already taken up so much time, you'll just have to wait longer. I'll have to see you again, after we get the radiology results."

The doctor was chastising me for disrupting the clinic's operational efficiency and his frustrated tone suggested that I had done something horrifying.

I sat awkwardly on the vinyl examination table, wanting to comply—to apologize, even, for my obstinacy. But I fundamentally disagreed with this doctor's practice of ordering X-rays on all his new patients before knowing anything about their symptoms. How many unnecessary scans were being performed in this clinic on the off chance of diagnosing bone cancer? What about radiation exposure and false positives that lead to more invasive testing and distress?

I thought back to my CT scan, when I'd felt uncomfortable, and had even signaled so with my questioning—only to finally comply with a test I didn't need. That wasn't going to happen this time. I did not want an X-ray, and despite the pressure, I was not going to let the doctor scare me into thinking that I had a high probability of something more sinister than inflammation.

"I'd prefer a physical therapy referral," I said. "If my pain doesn't improve in six weeks, I'll return for that X-ray."

I felt, even then, a strong urge to apologize and give in. But I didn't. I stood my ground. And when I walked out of the exam room that day, physical therapy referral in hand, I felt proud of myself. I had not caved to someone else's desires for me, just to avoid making a scene. It was liberating, to be true to my values in this way. And, after six weeks of successful physical therapy, my shoulder pain diminished. I never went back to that office again.

Picture It First

When we consider *What does a person like me do in a situation like this?* and connect to our *responsibility* to act defiantly, we then need the *ability* to defy. Our ability to defy improves with repetition and familiarity. The more we defy, or even just consider defying, the more comfortable and skilled we become doing it.

The first time we speak up, we might stumble, but with repetition our voice grows more confident.

Preparation and practice are key elements in our ability to surmount the final step into taking action. The Old Testament professor and the engineer, both from Milgram's experiments, felt not only a duty to defy but were also able to carry that duty out. Their confidence to act stemmed from their respective experiences, one from considering moral decisions that may harm others and the other from knowledge of the damage electrical shocks can inflict on humans. The more experience we have of situations in which defiance might be necessary, the more attuned we become to how our own tension manifests, and the better we get at progressing through the stages of defiance.

Experiencing situations in which we wanted to defy but didn't, or situations in which we did defy, regardless of whether or not it was successful, allows us to learn what factors may suppress our defiance and what factors enable it. We might engage in small acts of defiance to grow more comfortable. If we can just reach stage three of defiance—expressing our discomfort to the other—we are more likely to make it through the fourth and reach the fifth and final stage, an act of defiance. Like a weight lifter progressing from lighter weights to heavier ones, we can gradually build our ability, increasing our skills and confidence with each defiant act we undertake.

One way we can prepare and practice is by *anticipation*. When we are surprised by a novel situation, we often freeze, unsure of what to do, despite our intentions. Anticipating what we would do in situations that might require defiance can help us act quickly and in accordance with our values. Surprise disables defiance, while anticipation enables it.

Kevin, the police officer who defied an illegal order to search a garage, told me that one of the reasons he was comfortable taking a stand was because he had undergone his officer training shortly after the killing of George Floyd. As he learned about proper procedure and excessive force, he thought often about the rookie officers Alex Kueng and Thomas Lane who obeyed the orders of their former training officer—with tragic results.

He didn't want to be like them. He wanted to be the police officer who would not blindly follow orders and comply. He wanted to be the officer who would move the knee.

So often, we think of defiance as brave, but defiance can also be motivated by fear. The same fear that sometimes keeps us from being defiant can mobilize us to take action. Kevin feared another situation like the one that killed George Floyd, and he used that fear as a tool, a way of imagining how he could be defiant when the moment came. And he knew, as a police officer, it would come.

"I knew this job would be tough," he told me. "I knew I'd see unethical behavior."

He expected it. So he prepared for it.

Police training cannot only provide *awareness* of an officer's duty to intervene. Officers also need *behavioral* training in defiance to enact that knowledge. They need to role-play and practice it over and over. If they don't receive that type of training in the academy, they need to prepare themselves, as Kevin did.

When Kevin was ordered to search the garage, he imagined the worst-case scenario. He could envision the confused homeowner, jolted from bed, rushing out of his house with a gun, ready to confront whoever was in his garage. He could picture the resulting use of force from the police, justified by an imminent threat. He could foresee a situation in which an innocent man could be injured or even killed, simply because the police themselves were illegally searching his home. And it scared him.

Standing on the front lawn of that house in Oakland, watching as his older colleagues prepared to illegally search the garage, he acknowledged his tension, recognized that something was wrong, and then spoke up to vocalize his discomfort. When his concern was not taken seriously, he stated his wish to not comply, and finally defied the orders given to him. When the time came, he knew how to see beyond pleasing and obeying the immediate authority at hand by focusing instead on the broader consequences of his actions.

What happened that night, for Kevin, was bigger than an illegal garage search. It touched on his values as a police officer and as a person. He recognized that he was being asked to betray his values. He knew the law, and he knew the search was illegal. He recognized that it was the right time and place for him to defy and he connected with his values and responsibilities—even though doing so opened him up to isolation and shunning within his department, not to mention an official reprimand. He knew the risks, but he was ready.

That night, Kevin's defiance was not as consequential as some of the scenarios he might have imagined; nobody got hurt, the homeowner didn't come out guns blazing. But by enacting his values, he gained further experience and practice. His voice might have shaken as he told his lieutenant that the search had been wrong, and his childhood stutter might have slowed his speech—but by refusing to back down, he began to make a habit of not giving away his agency and power.

Next time, the stakes could be higher. But with each act of defiance, Kevin becomes more practiced in and comfortable with challenging others, more skilled at assessing his environment, and more able to predict how his actions could change it.

When we prepare and deliberate in advance, we can become better able to anticipate and recognize situations that call for defiance, eliminating the element of surprise that so often suppresses our tendencies to defy. Visualizing what our defiance might look like in a particular situation helps us rehearse what we might do or what we might say. A failed attempt often brings reflection on what we wish we had said. Pre-scripting and role playing have an important purpose. Practicing out loud before we are immersed in a difficult situation allows our mouths to get used to saying defiant words and our ears to get used to hearing them. Defiance starts to feel empowering and more like who we want to be.

Rehearsal and practice change the neural pathways in our brains—and these small shifts in our brains' wiring allow us to defy with less difficulty. These internal changes, which enable us to defy, subsequently change the trajectory of events that follow, leading to different outcomes. Even though we may be unaware of it, we often start to prepare for defiance months or even years before we put it into motion.

My experiences in those doctors' offices, a year apart, are a case study in an evolving ability to defy. We can get better at defiance with each and every experience, regardless of whether we defy or comply.

In the first instance, when I was urged to have the CT scan, I only reached the third stage of defiance. I had felt tension (stage one), and I acknowledged it to myself (stage two). I then attempted to escalate by vocalizing my discomfort to another (stage three). I asked clarifying questions, hoping the technician would hear the unease in my voice and register that I did not want to undergo the procedure.

But at the time, I couldn't go any further than that. I didn't know how to tell them I was not going to comply (stage four), so I could not reach the final act of defiance (stage five).

My failure to get beyond the third stage meant I went along with the unwanted scan.

This time at the orthopedic clinic, though, things were different. When I found myself in a similar situation, feeling the same tension (stage one) and acknowledging it for what it was (stage two), I escalated my discomfort more clearly to the medical assistant (stage three). I repeated that discomfort each time I was asked to submit to the X-ray. I pushed back forcefully, telling her I wanted to see the doctor first and would not comply with the X-ray until then (stage four). Then, when the doctor gave no legitimate reason for why an X-ray was necessary, even though I was uncomfortable rejecting his recommendation, I finally was able to give my informed refusal to the procedure; I was able to defy (stage five).

So what changed? Why could I refuse an unnecessary X-ray, when before I had acceded to the CT scan that I knew I didn't need?

During the year between those two doctors' visits, I had spent a lot of time thinking about why I simply went along with that CT scan when I didn't want to. Even if I didn't know it at the time, this reflection was training me for future defiance.

I had become deeply familiar with the tension I felt, the discomfort I experienced when being pressured to endure a procedure I did not feel put me, the patient, first. My unease made me reconnect with my core values about how medicine should be practiced within the four principles of healthcare ethics: putting patients first (beneficence), doing no harm (nonmaleficence), fairness in medical decisions (justice), and the right of the patient to choose (autonomy). I recognized that being coerced into taking unnecessary medical procedures violated all those principles—and by complying, I was doing nothing to change the situation for myself or for others.

Ever since that CT scan, I had learned to recognize my internal rationalizations for acting against my interests. My fear of being a "bad" or "disruptive" patient had previously kept me from enacting my values. Those rationalizations led me to reflect on who I am and what I stand for.

I had learned to rapidly assess medical environments and I could see clearly that the X-ray recommendation was less about me and more about the protocols of a profit-driven medical office that bills per procedure. I had evaluated that it was safe to defy and that saying no would be effective.

In my mind, I replayed my failed defiance during the CT scan, and I anticipated a similar situation. And in my head, I practiced what I wished I had said that first time. That anticipation, preparation, and practice changed my neural pathways and enabled me to act with determination when it was needed. I had a new sense of my ability to defy, and believed myself fully capable of carrying out a True No.

As a physician, I also understood that I was in a unique position not only to avoid a practice I found problematic, but also to potentially create positive change beyond my own needs. My refusal might just create a ripple that could have a positive effect for others. I felt responsible not only for my own health but also to advocate for all patients to receive better quality healthcare.

When I asked myself, *What does a person like me do in a situation like this?*, the answer had become simple and clear.

Creating Ripples

You might not think that an action as seemingly personal as refusing an X-ray would have much impact beyond my own life. Certainly the physician I met with did not immediately dismantle his office's patient intake practices as the result of my questions. But he was forced, in that moment at least, and however briefly, to consider how he treated new patients. Perhaps this reexamination introduced, for him, some sliver of doubt about how wise the automatic X-ray practice was. Or perhaps it didn't. But suppose another patient questioned his practice and also refused the X-ray. Maybe next time, the physician's doubt would grow and ultimately make him change his mind. Change often starts with just one person's defiance.

I didn't stop at my refusal. I told the story to friends and colleagues. They encouraged me to tell it more broadly—to outline the experience to other physicians in the hope it would bring about change. I then wrote a paper, titled "Investigations Before Examinations: 'This Is How We Practice Medicine Here,'" which was published in a widely read top medical journal. The editor added her own supportive article titled, "Testing Before Seeing the Patient," which linked to mine. These papers sparked national interest, and my inbox filled with messages from physicians across the country. Most congratulated me and highlighted many other examples of tests being automatically ordered before they could talk to their patients; others rationalized why such X-rays were necessary, efficient, and safe; and others lamented the system and emphasized it needed to change before they could. By writing about this personal, small moment of defiance, I had expanded its scope and started a larger discussion, one that resonated with others working in healthcare.

Even when it *seems* mundane, how we respond to small injustices that grate on our conscience can have ripple effects. Our True No might seem quiet in the moment, but it is rarely minor—and truthfully, neither are the injustices we are addressing. This is a fundamentally optimistic view of defiance; a recognition that when we challenge tiny tyrannies, we don't only help ourselves—we help others like us.

A colleague of mine whom I'll call Robert described a situation from his childhood that also illustrates the power of small acts of defiance. When he was in high school, it was relatively common for teenagers to use the word "gay" as a pejorative, to describe things they simply thought were stupid or uncool. This was the late nineties in the United States, in a relatively socially conservative and affluent part of south Texas, and although my colleague always felt uncomfortable with the way the word was tossed around, he never said anything to stop it.

Thoughtful and bookish, Robert was more interested in the debate club than the football team. With his shoulder-length hair and battered notebook full of half-finished poems, he fancied himself a liberal free-thinker in a sea of cowboy hats and traditional values. And although he felt perfectly comfortable arguing with his parents and classmates about the war in Iraq, he never said a word when his friends uttered a homophobic slur.

"I didn't know any LGBTQ+ people at the time," Robert said. "My friends didn't, either. At least as far as they knew. And so although I felt it was a little wrong, I didn't say anything."

At the start of the new school year, Robert made a new friend, a girl named Annie. She had just moved to Texas from San Francisco and had brought with her progressive social ideals that she was not afraid to broadcast. The first time she hung out with him around his friends, she immediately noticed the way they were using the word.

"Stop saying that," she said. "It's really not cool."

She hadn't raised her voice, but she didn't seem nervous about challenging her new friends, either. The group grew quiet.

"Come on, we're just joking around," one of Robert's friends said. "It's not that serious."

"I know you're joking around," Annie said, without a trace of hostility in her voice. "But it's not legit. So stop."

When Robert first heard Annie's protests, he was flustered, and embarrassed for his friends. He had never objected to such language as he simply dismissed it—he went along to get along.

But what happened next surprised him. The way Annie spoke up was so confident (and cool) that the group's language changed—almost immediately. Within a few days, "gay" as a synonym for "stupid" had vanished from the group's social lexicon whenever Annie was around—and over time, it was gone from the group even without Annie's presence, and then eventually from the broader school as well.

"When I think back on that," Robert told me, "it's hard to believe. No one really appreciated that kind of offensive speech. But we did tolerate it, until one person told us to stop."

The Defiance Domino Effect

Robert was so inspired by Annie's small defiance and its positive effects that he too now feels confident taking a stand when he hears derogatory language. Recently, when he heard his father using transphobic language, he challenged him just as Annie had once challenged him. When we take action, even in situations that may seem minor, the effects of our defiance often reverberate beyond the situation.

Part of what makes defiance so powerful is that its spirit is contagious; its success is not limited to individual events and actions. When we voice a True No, others often follow suit, even when the stakes are high and the risks considerable. This is how lasting change often occurs: through the accumulation of dozens or even hundreds of moments of small and not-so-small defiant acts. A True No's implications can reach far beyond the personal and into the realm of the political, the social, the global. We may create change in ways we could never have imagined.

Defiance's power is not as a momentary decision, to change one situation, but as a sustained practice that governs our life. A healthy defiance practice keeps us prepared for the crucial moments when we must take a stand. We need to develop our defiance muscle memory, because the situations in which defiance is most necessary are often those in which it is the hardest to think clearly.

A quote from Greek poet Archilochus, often attributed to Bruce Lee, highlights the importance of practice: "Under duress, we do not rise to our expectations, but fall to the level of our training." I believe that is true, not only in martial arts, but in defiance. Although we all have different backgrounds, and our own unique relationships with defiance, repeated and intentional practice can change us; it can rewire us for more consistent moral action and more effective defiance.

Neuroscientists have documented changes in our brain's prefrontal cortex when learning new tasks. Consistent training can literally change our brain's wiring. By building new neural pathways for defiance, it becomes easier and more accessible to us. At first, for those of us who find defiance difficult, it simply becomes more plausible, a new possibility. We may be clumsy at first and we may make mistakes. But over time, our focus improves, our thinking gets clearer, and our reactions become faster. Eventually it becomes more natural, more automatic, and an important part of who we are and what we stand for. Defiance, at last, becomes so ingrained that we don't even notice we're doing it. It is just us, being who we are.

14

You Don't Have to Be Brave

On August 20, 2018, the first day of the new school year, the teenager woke early.

She ate breakfast and packed her backpack. Putting her helmet over her pigtails, she wheeled her white bicycle out onto the street and set off into the heart of the capital. Her father would follow her, pedaling his own bike behind her.

It was a gorgeous morning, clear Nordic light slanting through the oaks along the river. Small for her fifteen years, with wide-set eyes and stern, elfin features, the teenager pedaled determinedly along the riverside cycle lanes, bypassing the rush hour motorists in their Volvos and Volkswagens. Steamboats bobbed in the water, belching thick black smoke that hung in the air like thunderclouds before dissipating. She sped over the bridge, accelerating on the downslope toward the heart of the city.

She hadn't ridden her bicycle much in recent years. The truth was, she had been ill for almost half of her life. There had been times she couldn't eat, times she couldn't speak, never mind pedal around the block. Today, though, with a renewed and resolute energy she was riding again. And she had something to say.

She glided through the city, tourists and shoppers already filling its streets, the high stone buildings casting enormous shadows over cobblestones and pavement. Ahead of her as she neared the bridge over the Söderström river was the Riksdag, the center of the Swedish government, sitting like a fortress on an island in the center of Stockholm. She coasted to a stop, and her father arrived shortly after. They locked her bike to a metal railing.

"Run along to school now!" he said, handing her a painted wooden sign that he had been carrying under his arm.

Passersby would not have heard this as the joke it was. In fact, they probably didn't even notice the teenager that day, nodding to her father and walking across the bridge and under the arch to the Riksdag. If they did, they probably thought she was one of the thousands of schoolchildren walking through Stockholm that morning, waving goodbye to their parents and fixing their hair after taking off their bicycle helmets.

But this teenager had a plan. She propped up the piece of wood against the parliament building wall. She'd bought it two weeks before at a building supply store for twenty kronor, then taken it home and painted her message onto it to be clearly legible from far away. In all capital letters, in black paint, it read:

SKOLSTREJK FÖR KLIMATET
SCHOOL STRIKE FOR THE CLIMATE

The girl opened her backpack, removed a blue cushion, and placed it next to the wall. She pulled out one hundred flyers, stacking them neatly and weighing them down with a stone, so that they wouldn't fly away in the wind. Then she sat on the cushion, her back propped against the granite wall, her knees drawn up to her chest.

All around her, people hurried to work, checking their watches and glancing down at their phones. She wondered when she would be noticed. She wondered if it would be an adult or a child who stopped first to read the flyers at her feet, their edges fluttering slightly in the breeze.

She had been careful getting the wording just right, because she knew it was probably the most important thing she'd ever said in her entire life. Here is a rough translation of the Swedish:

We children don't usually do
as you say.
We do as you do.
And because you grown-ups don't give a damn about my future,
neither do I.

My name is Greta and I'm in ninth grade.
And I am going on strike from school for the climate
until Election Day.

Greta Thunberg is now an iconic figure in climate change activism. Since her first school strike in 2018, she has addressed the United Nations, given speeches around the globe, and become one of the world's most recognizable faces in the fight against climate change. Whatever your beliefs about Thunberg or global warming, it is undeniable that her words and actions have inspired millions to rethink their views.

People often use words like brave and courageous to describe her, in no small part because of the obstacles she has had to overcome in speaking out.

"I was diagnosed with Asperger's syndrome, OCD [obsessive-compulsive disorder], and selective mutism," she said in 2018. "That basically means I only speak when I think it's necessary. Now is one of those moments."

But it's not merely bravery that defines Thunberg's defiance, but something much deeper.

She knows exactly who she is.

Becoming a Moral Maverick

One especially widespread and unexamined belief about defiance is that it requires bravery, and that strength is a prerequisite. People who go against the flow are brave, the logic goes. Those who speak truth to power are strong.

That can be true, but defiance is not reducible to strength or weakness, courage or cowardice. It is not solely for the brave, the strong, or the extraordinary. We all have the capacity to be defiant. There are many paths to defiance, but they all begin in the same place: one person, connecting with their core values.

Greta Thunberg connected with hers early in life. Sitting in a Scandinavian classroom at age eight, she watched a film of melting ice caps, starving polar bears, and plastic in the ocean. Driven by fear, she searched to know more. She devoured books about the climate crisis and began following the research of climate scientists and activists. As she grew older, she gained enough knowledge and understanding to know that she could not consent to a world in which climate change was not addressed and combatted.

Starting with her parents, she attempted to educate and inform them, pressing on them book after book, documentary after documentary, study after study. And it worked. Her father was so moved by what Thunberg showed him that he became a vegan. Both her parents made the decision to forgo airplane travel, despite Thunberg's mother's career as a touring singer.

Greta Thunberg then moved on to educating others. When she sat down outside the Riksdag that day in August 2018, her public defiance was the product of preparation and—crucially—introspection. When she asked herself, *Who am I?*, she knew the answer. She knew what a person like her should do in the situation she found herself in.

·   ·   ·

Through our training, we can fundamentally shift our relationship with defiance. We can transform ourselves from credulous conformists into *moral mavericks:* people who use their Defiance Compass and are prepared to go against the grain in service of their most authentic values. Moral mavericks speak up, speak out, and demand change when necessary.

When we endeavor to become moral mavericks, we cease to be spectators in our own lives, and become disrupters of the status quo. We reclaim our agency: Rather than focusing on what is done *to* us, we focus on what we can *do*.

Moral mavericks are proactive, not reactive; conscientious, not impulsive. They do not airily dismiss the values of the societies they live in or even organizations they belong to. They are not jaded or cynical about others—quite the opposite. They connect with their moral values and avow responsibility, so much so that—like Sara, the grants coordinator, or Rachael Denhollander, the former gymnast—they will fight the institutions they love for not living up to those stated values.

Moral mavericks understand that small-scale defiance is still defiance—that the accumulation of these small acts of defiance makes a person who they are. And more than that, they understand that "small" defiance often grows.

When we think of ourselves as moral mavericks, we will answer the last question in the Defiance Compass—*What does a person like me do in a situation like this?*—with principled action. The situation could be life-or-death, as it was for Angeli Rose Gomez, sprinting toward the elementary school where her young sons were hiding from a gunman. Or it could be more quotidian, as it was for me in my doctor's office.

Becoming a moral maverick is not a switch that one flips, but a lifelong dedication to the practice of defiance. It's not about immediate transformation and getting it right every time; rather, it's about progression, and understanding what factors help us, what hinders us, and how we can overcome the hurdles that might trip us up in the future when we need to defy.

Past experience can guide us. The process of becoming a moral maverick involves examining the times we felt tension and let it slide, only to feel worse afterward. It encourages us to look anew at the social norms and psychological constraints that keep us compliant and silent, and to make efforts to understand how to safely oppose them in the future. It pushes us to be comfortable making ourselves uncomfortable, when it is required to honor our principles. A moral maverick recognizes the ways in which they have lived as a conformist—and then endeavors to do things differently.

When you become a moral maverick, you change yourself, your surroundings, and maybe even the world.

Forging Our Own Path

Many people are familiar with Malala Yousafzai, the Nobel Prize–winning Pakistani activist for girls' education. Fewer know the name of her father, Ziauddin Yousafzai, an activist and educator in the Swat Valley of Pakistan. Throughout her childhood, Malala watched her father give speeches, raise money, and work tirelessly to promote children's right to education and peace through active pacifism. A believer in freedom, democracy, and open expression, he allowed his daughter to sit with the men as they debated politics, encouraging her to speak up for herself and to value her education as much as he had valued his own.

"You are a child," he always told her, "and it's your right to speak."

At age eleven, she was already captivating the audience of BBC Urdu with her anonymous diary of life under Taliban rule, and on Pakistani TV she once asked, "How dare the Taliban take away my basic right to education?"

We often learn to be defiant from role models or significant examples we experience in the course of our lives. Ziauddin was a strong exemplar for Malala. Rosa Parks noted the defiance of her mother, Leona McCauley, refusing to move on the bus years before Parks herself did the same.

Not everyone grows up with a father like Ziauddin Yousafzai or a mother like Leona McCauley. Many of us lack a consistent role model of defiance. We piece together what a True No looks like, building an image of defiance out of bits and pieces of resistance and protest that we encounter. For myself, models of defiance have come intermittently. I remember them as scenes that stood out from everyday life, remarkable for their incongruity: my mother standing up to the bullying boys in the alley, my father arguing against the town council about the injustice of a denied bus pass.

But even without explicit guidance, we are all capable of becoming moral mavericks. Greta Thunberg's path to defiance was self-directed. Although her parents supported her, they did not spark her activism. In fact, it was the other way around: Thunberg educated them about the perils of global warming. That is not to say she did not receive guidance. Indeed, it was the Parkland survivors—American teenagers who coordinated a national student walkout to protest lax gun laws, after a mass shooting at their school—who inspired Thunberg to organize her own school strike.

All of us can find our own path to a True No, because all of us can learn to connect with our core values. What unites moral mavericks is not upbringing or role models but self-knowledge. It is not circumstance that defines our defiance, but our own inner self, our connection to who we truly are and what we truly value.

We often think of defiance as a loud and public gesture. The word itself often conjures up iconic images: David swinging his slingshot beneath the mighty Goliath; Tommie Smith and John Carlos raising their fists in a Black Power salute during the 1968 Olympics; the anonymous "Tank Man" of 1989's pro-democracy protests in Tiananmen Square, staring down a convoy of tanks. Such public displays of defiance have become shorthand for a kind of almost unimaginable bravery in the face of danger, oppression, and violence.

But—as we've seen—so much of defiance happens behind the scenes. The images described above are the culminating events of a longer process of defiance, which happens individually, privately, and quietly.

Greta Thunberg is now a defiant figurehead for a large-scale global movement. But what inspires me about her defiance is her inner process—all those months and years climbing out of the depths of depression, and into communion with the beating heart of who she is, and who she aspires to be. It is not the image of her wooden sign that represents the enduring power of her defiance; instead, what I think of are the laborious hours of planning her protest, deciding what to paint on her sign, and the even more prolonged work of preparing what to say.

There were moments, early in the school strikes, when Thunberg seemed overwhelmed. Once, a few days after she first sat down with her sign, a group of elementary school children started to approach her at her usual spot at the Riksdag, and Thunberg—who had been relentlessly bullied for being "different" in school at that same age—panicked. She began to cry and stepped away.

But then she paused, collected herself, and walked back to the children to explain what she was doing.

Greta Thunberg is not an icon but a human, bruised by a childhood of illness, bullying, and pain. She defied not in spite of these things, but in some sense, because of them. They made her who she is.

Greta Thunberg, Jeffrey Wigand, Rosa Parks, and Malala Yousafzai are all moral mavericks. Their actions are deeply considered expressions of their true values. They defied because it was the only way they could live with themselves. When we know who we are and what type of world we want to live in, defiance can make us into the best versions of ourselves.

There are many Thunbergs, Wigands, Parkses, and Yousafzais in this world who do not achieve renown, their actions witnessed only by friends and family. They do not appear on television or in large conventions. While their stories might go unnoticed by the wider world, that does not mean they go unfelt. Small moments of defiance can add up to larger epochs of resistance; tiny acts of *no* can spark revolutions. Defiance doesn't only topple dictatorships, overturn repressive social orders, or inspire broad social change. It also helps bring us into connection with our true selves.

Defiance can be brave, but it does not require world-renowned bravery. Defiance can lead to extraordinary change, but it does not stem from only extraordinary people. Defiance simply means knowing exactly who you are and acting in alignment with those values. And that itself can create great change.

"I'm gonna do something; I'm gonna go [on a] school strike," Greta Thunberg said later. "I didn't really think this might lead to a global movement, but it did."

Breaking All the Rules

It may be difficult, at least initially, to think of yourself as a moral maverick. It can be challenging to see through your wiring for compliance, created by years of social conditioning, culture, and upbringing. Even for seasoned moral mavericks, defiance can feel impolite or awkward. Few of us are immune to the pressures around us. Rationalizations and self-deceptions afflict us all, and it can sometimes be more comfortable to relinquish our responsibility to authority than oppose it.

Even when you feel the stirrings of defiance, the forces opposing your autonomy can seem insurmountable, causing you to lose sight of your actions' potentially broader consequences. Over time, continued conscious compliance might even mute your internal voice of defiance.

But that voice is rarely completely extinguished, because it is at the core of who you are—of who we all are. Even if you have spent most of your life coloring inside the lines, following the rules, and obeying the status quo, it is always possible to reach your inner moral maverick—to find your voice and activate your potential for defiance.

It can take years, and it is not always easy. Along the way, there may be pushback and setbacks. It must be learned, practiced, and learned again.

I know this all too well. Indoctrinated as a child with the belief that obedience was good, I had to learn how to defy in ways that were not merely imitative but true to myself. As a young adult, I often felt torn between obedience and open defiance, and even after I had gained a more nuanced understanding of how I had been conditioned, I vacillated between outward compliance and inner rebellion.

A few years ago, a junior job candidate I was interviewing asked me something genuinely surprising as we walked across campus.

"How do you break all the rules?"

"What do you mean?" I asked.

I'd thought that if anything, I'd played too much by the rules, doing what was asked of me and what I felt was necessary to obtain my academic tenure—working long hours, doing research, teaching different courses, going above and beyond expectations with service and committees, and on occasion deploying my crocodile smile. I certainly didn't think my path had been outwardly defiant.

"Well, you're at a business school," Becca, the job candidate, answered. "But you don't just publish in management journals. You're putting out work in medical, psychology, law, and ethics journals. We are told not to do that!"

"I hadn't thought of it that way," I told her. But as we cut across the grass to get her to the next meeting on time, I realized she was right. I had been following my own path. I knew the expectations in my field, I knew the sort of publications that we are rewarded for at business schools—where my career has taken me—and I had actually been warned to steer some of my work away from medical journals and toward management ones. But I had always wanted to publish papers in the places where my research would have the most impact. I hadn't been guided by what I was *supposed* to do, but by what I thought could create the most positive change. I wanted my work to make a difference in the world. By simply staying true to myself, perhaps I had "broken all the rules." And I hadn't even realized it.

Sometimes we jump from acknowledging our tension (stages one and two of defiance) right to defying (stage five of defiance). Sometimes we skip the stages in between or move through them so fast, we don't register them. Sometimes this behavior feels so natural, so right, and so attuned with who we are, that it may not even appear to be defiant to us. When we break rules to remain true to our deeply held values, we are being defiant, and that defiance can register with—and inspire—other people even when it does not register to us.

This, in a way, is the defiance dream: living in such alignment with your values that you act on them seamlessly, without hesitation. Defiance becomes not just a response to crisis, but a natural way of being in the world.

Our Aspirational Self

Defiance like Greta Thunberg's can have durable effects on the world. But just because it is focused outward does not mean it has no internal effects.

In interviews, Thunberg has been open about how she views her neurodivergence as her "superpower," one that helps her focus on what truly matters and act in accordance with her core values.

She reconceived what others—including even her parents—might have interpreted as a liability and turned it into a strength. Her defiance evolved in tandem with her understanding of how her brain worked. It allowed her to understand herself and how she could be most effective at enacting her values to protect the world.

As she put it once in an interview:

> It feels like many today—neurotypical people, people in general—are so focused on following the stream, doing like everyone else, because they don't want to stand out. They don't want to be uncomfortable. They don't want to cause any problems. They just want to be like everyone else. And I think that's very harmful in an emergency where we are social animals. We're herd animals. In an emergency, someone needs to say that we're heading toward the cliff.

Thunberg thinks of herself as one of those voices in the emergency. She has long known that she is different, and she uses that difference—leaning into, not away from, the aspects of herself that are not like everyone else.

Defiance allows us to connect with our aspirational self, to become who we want to be. We do this not by rejecting the parts of ourselves that make us unique but by embracing them. When we embrace what makes us special, makes us different, we can defy in new ways.

You don't have to change who you are to be defiant. You have to become *more* yourself.

This insight allows defiance to become a part of our everyday life, lessening the tension so that it feels right and becomes routine, even commonplace. Repetition eliminates the static, the uncertainty, and the anxiety of defiance, so that we can focus on who we aspire to be and what to do. Models and guides can be helpful as we learn to defy. Ultimately, though, a moral maverick finds the lodestar of their defiance deep within themselves.

After my conversation with Becca, the job candidate, about "breaking the rules," it dawned on me that what made me different was what made me able to defy. Rather than seeing myself as someone who struggled with defiance—as I had for many years—I began to realize that my approach to my career had actually been a long-running signal of defiance to others. This aspect of my life that made me unique could be a way to circumvent my wiring for compliance, to tap into the places where I already had the power to defy, then derive strength, clarity, and lessons from it.

To become defiant, I just needed to become more of who I already was.

Becca's comment broke my self-concept of always being a compliant person and helped me to find the rebel yell in my everyday voice. When I learned to be true to myself, to lean into the qualities that made my defiance unique, I became *more myself.* I learned that a life of defiance is a more honest life, and in that honesty lies immense power.

Some of us defy by marching, holding signs and banners high, and some of us defy by speaking up to bosses and managers or to our friends and family members. Others express their True No through research or education, or through politics or art. However we decide to live in alignment with our values, we become part of a defiant collective.

Defiance is not a monolith, nor merely an expression of the will and power of the crowd. It does not march in complete lockstep. It is more of a song: a chorus of diverse voices, each singing their True No, some booming and loud, some soft and quiet. They are not always in tune with one another, not always singing from the same sheet, but they all, in their own unique way, perform key parts in the same defiant anthem. Thought of this way, defiance is not only a means to understand and alter the world for the better, but a pathway to reach our aspirational self.

Defiance changes our homes, our workplaces, our communities, and the world beyond. Yet one of its greatest gifts is that by enabling us to become more truly ourselves, defiance also changes *us*.

CONCLUSION

Dare to Defy

Sometimes, it is in our darkest of days, when we are standing in the shadows of oppression and death, that the spark to defy is born. I often think back to that day in May 2020, when George Floyd took his last breath in front of Cup Foods. I recall the small crowd that gathered on the street that day. There weren't many there to see the dying man's last moments, only a dozen or so. One of them, a high school student named Darnella Frazier, filmed what she saw.

Frazier did not plan to start a social justice movement. She was just headed to Cup Foods for some snacks with her nine-year-old cousin, Judeah Reynolds, in tow. You can see both of them in the police body camera footage: Frazier in flip-flops and blue sweatpants, her long black hair hidden beneath her gray hoodie, gripping a cellphone in a yellow case, Reynolds in black Nike slides and leggings, holding a bag of candy and wearing a teal T-shirt with the word "LOVE" printed in bold lettering. The expressions on their young faces move me deeply: Frazier looks worried and shocked, her head tilted upward, her mouth slightly open. Reynolds is looking up at her older cousin, as though seeking reassurance.

When Frazier saw what was happening, she took out her phone and told Reynolds to go back inside the convenience store. For the next ten minutes, she filmed the scene: Derek Chauvin kneeling on George Floyd's neck, Alex Kueng holding his back, Thomas Lane holding his legs. She kept filming, even though the fourth officer at the scene, Tou Thao, told her to stop. In her video, you can hear the crowd yelling at the police to stop and pleading with them to check George Floyd's pulse.

That same evening, she posted her ten-minute video online. Within hours, it had gone viral, and Floyd's death became the center of a nationwide reckoning. Hundreds of thousands took to the streets in cities across the United States and around the world that summer, demanding justice and police reform. When the four officers involved were tried for murder, Frazier's video was a foundational piece of evidence for the prosecution. If we know the true story of what happened that day, it is largely because of Darnella Frazier, who refused to put her camera away even as an officer barked at her to stop and Derek Chauvin reached for his Mace.

"It wasn't right," she later said at Chauvin's trial.

Frazier and the other bystanders may have felt helpless in that moment, but as she proved, there is a lot of power in being defiant, even as a witness.

She later wrote that George Floyd was not the first person she'd seen killed by the police, but he was the first who had been killed right in front of her. "A lot of people call me a hero even though I don't see myself as one," she wrote. "I was just in the right place at the right time."

Darnella Frazier was in the right place and time for *her*. She knew that what she was seeing was wrong. Her young cousin, Judeah Reynolds, also recognized the cruelty and shouted out, "Get off of him." Frazier felt a responsibility to do something about the injustice in front of her. She was prepared to defy.

Each one of us, at some point in our lives, will witness something unjust, be it in a McDonald's office or a tobacco company laboratory, during a police response on the street, in a conference room at work, or with our own family around the dinner table. Sometimes we are the victims of that injustice, and sometimes we are bystanders, but either way, we are often caught between wanting to act and being unsure of exactly how to go about it. How might things have gone differently if Alex Kueng, Thomas Lane, or any of the other officers present had been appropriately encouraged and substantially trained to defy when needed? Would George Floyd still be alive?

More broadly, what does a society that makes space for defiance look like? What would our world be like if we all felt the power to say no?

Situations that require defiance occur over and over, every day—but they don't *have* to end the same way. Police officers *will* get asked to comply; employees *will* be encouraged to look the other way; women, people of color, and other marginalized groups *will* be pushed to swallow their objections to slights and intimidations. But we can start preparing right *now* to handle those inevitable situations. Knowing who we are, assessing the situation, taking responsibility, and increasing our ability with preparation and practice rewires our automatic responses so we can do what we once deemed impossible.

I have come to recognize and welcome dissent, both my own and that of others. It turns out that my career path is more defiant than I realized, and I've become more comfortable asking questions in situations where routine compliance is expected. I encourage it in my students, my colleagues, and even in my adolescent son. Rather than teaching him either to comply or to go with knee-jerk rebellion, I am encouraging him to explore his own values and to know his own True Yes and True No. I want to raise a moral maverick, someone who complies with policies and structures that are in line with their moral code—but is willing to reject any structure that isn't. When my son sees something—to paraphrase the famous NYC subway poster—I want him to *say something*.

Darnella certainly did. The experience changed her. "It changed how I viewed life," she wrote.

Watching that video, and thinking of her actions, changed my own.

And it changed the lives of so many others.

It changed how someone like Kevin, the police officer, viewed his training and work. He considered his aspirational self—how would he want to respond in the rookie officers' shoes? He started to anticipate, prepare, and practice for such moments as he knew they would surely arrive. And when one did, he was ready and expressed his True No.

From such personal change, much structural change is possible. Moral mavericks rarely overthrow systems by themselves. But revolutions can start with something as small as the click of a phone camera, a shake of the head, a quiet voice saying, *No—no, that isn't right.*

You have to train to be a moral leader, a moral maverick. The transition doesn't happen instantly. But with practice, it becomes easier. After a while, and through repetition, our muscles for defiance grow stronger. Our decisions become clearer, and we develop more skills. Defiance starts to become more natural.

Rosa Parks hadn't resisted in that way on the bus before. Neither had Greta Thunberg before her first school strike. Neither had Darnella Frazier. But what happened before their acts of defiance—their experiences, their thoughts, their deliberations—was critical for all of them.

When Rosa Parks said no on that Montgomery bus, what world was she fighting for? When Greta Thunberg sat down in front of her country's parliament, what was she reminding the adults around her to see? When Darnella Frazier took out her cell phone, what community did she imagine for her cousin Judeah? For herself?

Defiance is not just saying no to something that is not right. It's saying yes to the opposite, to the world you want to create.

As someone who has come to value and practice defiance, I expect if someone asked me, *What is the one thing you do differently now?,* I would answer: I give myself the power of the pause.

I keep with me the quote of Austrian psychiatrist Viktor Frankl, who wrote *Man's Search for Meaning,* as a perfect way to describe the moment before we choose to defy:

"Between stimulus and response, there is a space. In that space is our power to choose our response. In our response lies our growth and our freedom."

Take the space. Take a pause.

Sometimes it's a moment, sometimes it's a day, but take the time to ask yourself:

What does a person like me do in a situation like this?

Pause.

Sometimes the situation is clear, and the answer comes quickly. Other times, the stakes are higher; there are nuances, and consequences to consider.

Every situation that might call for defiance—or consent—is different. But now, I aim to stop to consider my response instead of simply falling into what's expected.

Defiance is a choice. Your choice. That in itself is incredibly liberating—not to mention powerful.

Being a moral maverick is knowing a new way of saying no. Rosa Parks, Malala Yousafzai, Jeffrey Wigand, and Greta Thunberg changed history by doing so. So did Colin Kaepernick and Michael Quinn, Angeli Rose Gomez and Rachael Denhollander.

Malena Ernman, Greta Thunberg's mother, put it this way:

> It's been said that the moment a person acting alone is joined by someone else, a movement begins.

Reading those words reminds me of Michael Quinn, the police officer from Minneapolis who broke through the blue wall of silence and wrote, "All it takes is one."

One like Kevin, the police officer who refused to search that Bay Area garage and who believes that a new era of policing is on its way.

One like Jeffrey Wigand, who publicized the harmful and addictive effects of nicotine.

One like Rosa Parks, who refused to move.

One like Greta Thunberg, who will not let apathy on climate change prevail.

One like Malala Yousafzai, who wants every girl to be able to go to school.

One like Darnella Frazier, who refused to look away from injustice and death.

One kid in the alley, who stands up to the others, so that an immigrant mother doesn't have to put them in their place.

One person like you.

For far too long, compliance has been our default. We have said yes so often, and for so long, that we have forgotten what it means. And we have reserved *no* for the iconoclasts, the revolutionaries, and the troublemakers.

Now more than ever, in our schools, in our homes, on our streets, and in the halls of power, we need moral mavericks. We need to encourage them, foster them, celebrate them, and *become* them.

It is time to put *no* on an equal footing with *yes*. To imagine a world in which unjust orders are questioned, unfair structures interrogated, and flawed assumptions overturned. Not out of petulance but out of genuine thoughtful consideration. A world in which we only say yes to the things that truly align with our core values.

Such a world, built on a foundation of earnestly examined individual choices, is a more ethical one, a less violent, less exploitative, and freer place for everyone.

A world in which yes is not taken for granted, and no is not discouraged, is a world based not on compliance, but true consent.

It is high time we imagine it.

It is high time we create it: choice by choice, decision by decision.

Yes by yes.

No by no.

We create a world in which all of us can dare to defy.

Reader's Guide

A CONVERSATION WITH
DR. SUNITA SAH

1. What compelled you to write this book?

As a child, I believed what adults taught me: that being obedient was "good," that authority figures were to be trusted, and that I should do as I was told and not make a fuss. But in school, I saw firsthand the way teachers could abuse their authority, physically assaulting "bad" students in the name of discipline or order. Later, as a college student, I read Stanley Milgram's landmark study of obedience—in which many subjects complied with orders to administer what they believed were life-threatening electric shocks to another person. And I began to question what it meant to be "good," and whether obedience really was something we should always strive for and celebrate.

George Floyd's murder in 2020 was another turning point for me. That day, I saw echoes of Milgram's subjects from the 1960s in the rookie police officers who complied with the orders of their training officer. I realized that despite decades of historical lessons and research on influence, our propensity for obedience is still as strong and dangerous as ever.

I have always been passionate about fairness and justice, and the need to build an equitable world. Yet I have also always believed myself to be a compliant person. Throughout my life, the pressure to be "good" and obedient has often conflicted with my need to act in accordance with the values I hold most dear.

What I wanted, I realized, was to write a book that addressed that tension I felt. It would be the book I wished I'd had growing up: a handbook for defiance, a step-by-step strategy for learning how to defy when it matters most.

2. How is this book different from other books on how to say no?

This book draws from behavioral science. It is informed by my research and that of other scholars and offers a novel definition of defiance as a proactive *positive* force in our lives. Rather than viewing defiance as a single irrevocable act, which can be intimidating and prohibitive for those who need to defy, I break down the stages of defiance into five actions that are accessible to us all.

Defiance, in the popular imagination, is often seen as an attribute of extraordinary people who do courageous things. And certainly, within these pages, there are thrilling stories of people who defy. But a central concept of my book is that defiance doesn't have to be loud to be effective—in fact, it usually isn't. Defiance can be quiet and subtle, while still allowing us to move toward a better life, one that is more aligned with our values.

Being defiant is much more than just saying *no*. It is actually more about saying *yes*—to our values, to what we truly believe. My aim is to encourage readers to anchor their personal values in a broader social context: to examine how our everyday acts of compliance and dissent add up to the society we live in.

And as I've learned, even those small acts of defiance can make a big difference—in our lives and in society at large.

3. What do you hope this book does for readers?

I want this book to cultivate and nurture defiance in its readers—to show them that defiance doesn't have to be loud, it doesn't have to be oppositional, and most importantly, it is not the sole preserve of the larger-than-life heroes of history.

We don't even have to think of ourselves as "defiant" to access the power of a True No. That's certainly been true for me. Even now, after all my research and practice, I don't view myself as a person who can loudly and publicly defy. But I have learned to connect my values and my behavior more often, and I have realized I am defiant in my own unique way.

I hope this book will give readers a clear way to apply big principles to the smallest details of their lives. So many of us freeze in situations that don't feel right. This book is designed to make defiance accessible to all, to recognize that we have more options than the automatic compliance often expected from us. It gives us an action plan for defiance, one that starts long before a moment of crisis.

Beyond personal transformation, this book is a call to action for a cultural shift. I envision a world where defiance is not just accepted but encouraged in our children, our friends, our family members, and in ourselves. A world inhabited by what I term "moral mavericks"—people who, guided by their core values, can embrace their responsibility and speak up, speak out, and demand change when necessary.

Ultimately, this book is more than just a guide; it's a manifesto for the right to resist. My hope is to create a world in which everyone is free to defy.

4. Who is this book for?

This book is for anyone who grew up being told to be "good," to be polite, to not make a scene, and to do as they were told. It's for those of us who have felt our voices silenced, who regret not taking action, who wonder why we did something that we did not agree with.

It is for anyone who has ever swallowed their objections, apologized for being themselves, pasted a false smile on their face, and said, *Okay*—even though it went against everything they valued and believed in.

It's for people who wonder why they stood by, why they couldn't help.

It is for any of us who have been told that the world is not ours, who feel discriminated against or who have observed injustice, bias, sexism, racism, and bigotry. It is for anyone interested in empowerment and self-improvement.

It's for parents teaching their kids to stick up for themselves and others. For workers learning to stand up to their bosses and question unfair or harmful policies. For activists, to understand what drives them and to avoid falling for false defiance. For managers, who want to make space for constructive dissent from their employees, and for policymakers who want to ensure that when we're asked to sign our consent it really is our informed and freely given True Yes.

So many of us struggle with defiance. We worry that we'll make the wrong choice. We worry that we'll get hurt. We worry that even if we defy, nothing will come of it—it won't matter. The people profiled in this book—students and teachers, police officers and peace activists, professional athletes and nurses, engineers and managers, teenagers and parents—represent a veritable melting pot of occupations, races, cultures, ages, and life experiences.

Like defiance, this book is for everyone. We all need an education in how to enact our values under pressure. That's what I've aimed to offer here. The research and experiences captured in this book have changed my life; my hope is that it will change yours, too.

GLOSSARY

Compliance: Reactive or passive obedience, largely dictated and imposed by the wishes of others.

Conscious Compliance: The active, considered decision to comply in a situation in which you would rather not, because defiance would be overly dangerous, too costly, or ineffective.

Consent: A thoroughly considered authorization that is an active expression of our deeply held values. Consent requires capacity, knowledge, understanding, freedom, and authorization. It comes from within us and represents our True Yes.

Defiance: A thoroughly considered refusal that is an active expression of our deeply held values. Like consent, defiance requires capacity, knowledge, understanding, freedom, and authorization. It comes from within us and represents our True No.

Defiance Empathy Gap: The discrepancy between people's beliefs regarding the safety, ease, and effectiveness of defiance in a particular environment.

Defiance Hierarchy: A social order of who is allowed to defy and who is expected to comply, governed by social norms, stereotypes, and societal and cultural expectations.

False Defiance: Rebellion, resistance, or disobedience that stems not from our own values but from the values, wishes, or desires of another person, institution, or group.

Insinuation Anxiety: The concern, worry, or apprehension we feel about signaling a negative evaluation about another person to that person.

Moral Maverick: Someone who acts on their core values, realizes their responsibility to defy, and gives their True Yes or their True No in an effort to create a better world.

Quiet Defiance: Acts of defiance without vocal public declaration, often done secretly to avoid drawing attention to the defiant act.

Resistance to Resistance: A tension that paradoxically counteracts our propensity to defy.

Sales Pitch Effect: The pressure we feel to avoid appearing unhelpful, uncharitable, or uncooperative to another person.

Self-Connection: A process of introspection that leads to understanding who you are: your core values and highest principles.

APPENDIX

Chapter 1: Wired to Comply

Many of us are conditioned into thinking about defiance as a simple binary:

- Compliance = Good

- Defiance = Bad

Breaking this binary requires that we see defiance as neither good nor bad in itself, but simply as a way for us to access our truest self and act on our core values.

Chapter 2: Tension Is Your Strength

Resistance to resistance often manifests as tension: a deep discomfort arising when we find ourselves in a situation that calls for compliance, even though we would prefer to defy. We may experience it as nervousness, anxiety, or doubt.

When you feel tension, don't discount it. Instead:

- Examine it. How does it manifest? Is there a physiological component?

- Ask yourself: What is causing me to feel this way?

- Am I living up to my values or compromising them?

- Am I uncomfortable because I want to defy?

Tension is not weakness. It is strength: your brain and body's way of telling you that defiance might be necessary.

Chapter 3: Know Your True Yes

Compliance and consent are two different things.

> **Compliance** is reactive or passive obedience, largely dictated and imposed by the wishes of others.

> **Consent** is a thoroughly considered authorization that is an active expression of our deeply held values.

For consent to be valid, it must satisfy five elements:

1. *Capacity*—Do I have the competence and cognitive ability to make this decision? Am I impaired in any way?

2. *Knowledge*—Do I have the information and facts I need to make this decision? Do I know the risks, benefits, and alternatives?

3. *Understanding*—Have I achieved an adequate understanding of the facts?

4. *Freedom*—Am I being coerced? Do I have the choice to say no?

5. *Authorization*—Does this decision align with my true values? If yes, give your informed consent, your True Yes. If not, give your informed refusal, your True No.

Chapter 4: Break Free from Influence

Factors that can limit our ability to defy include:

- **Insinuation Anxiety:** the concern, worry, or apprehension we feel about signaling a negative evaluation about another person to that person.

- **Sales Pitch Effect:** the pressure we feel to avoid appearing unhelpful, uncharitable, or uncooperative to another person.

Within each of us there are two ideals, often in conflict:

- The *independent self,* which values autonomy, freedom, and individuality above all other considerations.

- The *interdependent self,* which prioritizes interpersonal harmony.

Balancing these competing ideals is an ongoing process, especially when it comes to defiance, where the social pressures to comply can be quite powerful.

Chapter 5: Reclaim Your Power

To lessen the social pressures keeping us compliant, get a little distance:

- *Physical* distance can insulate you from insinuation anxiety and the sales pitch effect.

  - Step out of the room.

  - Ask for a day to think things over.

  - Whatever you do, try not to make a decision in the moment.

- *Psychological* distance can also protect you from social pressures to comply.

  - Name what you're feeling—is it insinuation anxiety? Is it the sales pitch effect?

  - Speak to yourself in the third person. Ask yourself:

    – Does [say your name] really want to do this?

    – Is this what [say your name] really wants, or does [say your name] simply feel pressure to comply?

Chapter 6: Find Your True No

A True No stems from connection with our core moral values such as shared humanity, equality, and fairness.

Make sure you have the *capacity, knowledge, understanding,* and *freedom* to *authorize* a True No.

Prepare and practice for defiance by understanding the following stages of defiance.

FIVE STAGES OF DEFIANCE

STAGE 1	STAGE 2	STAGE 3	STAGE 4	STAGE 5
Tension	Acknowledgment (to ourselves)	Escalation (Vocalize to others)	Threat of non-compliance	Act of defiance

1. *Tension,* which can manifest cognitively, emotionally, or physically in the body

2. *Acknowledgment* of the tension and a consideration of why you feel it

3. *Escalation* of your concern to others

4. *Statement* that you will not comply

5. An act of *defiance*

You don't necessarily have to proceed through each of these stages in order for defiance to be possible. But understanding the progression from tension to action can allow you to reflect on your own journey to defiance—and to reassess, at each stage, how you wish to proceed.

Defiant action becomes much more likely after stage three: Once a person has decided to escalate their concern to others, it becomes more likely you'll progress to stage five.

Chapter 7: The False Defiance Trap

False defiance might *look* like defiance, but it does not arise from our true core values. Often, it is the product of knee-jerk opposition, moral convictions, or following the crowd.

Moral convictions can lead us to disregard the pressures and conventions keeping us from defiance—but they can also lead us to act in destructive and harmful ways that might violate our own values.

True defiance derives from our core values and highest principles, not our allegiance to a specific group, political party, or religious dogma.

Chapter 8: Who Gets to Defy?

In our world, some people are "allowed" to defy while others face harsher consequences for noncompliance. Social norms, conventions, and policies enforce a **defiance hierarchy** that mirrors larger divisions and inequities.

Conscious compliance helps us survive certain situations in which the costs of defiance might outweigh the benefits. Ask yourself:

- What are the costs of defiance—physical, social, psychological, financial, health-related—in this situation?

- Could defiance lead to positive change?

- Could the change outweigh the risks?

Conscious compliance is most effective as a short-term survival strategy. Remember that just as there are costs to defiance, there are also costs to continued conscious compliance.

Chapter 9: Quiet Defiance

Quiet defiance is an act of defiance without vocal public declaration. It may work in situations that might be dangerous or in which the costs of compliance may be too high to bear.

A road map for **escalating defiance in institutions** can guide us to affect change.

ESCALATING DEFIANCE IN INSTITUTIONS

Advocacy   Dissent   Disobedience   Principled Exit

1. *Advocacy*—enacting your professional values as part of your regular duties

2. *Dissent*—can be internal or public, and involves questioning particular actions or requests that do not align with your core moral values

3. *Disobedience*—openly disobeying rules or laws that are causing clear harm

4. *Principled Exit*—leaving a situation, environment, or organization when its values do not align with your own

Chapter 10: Who Am I?

When considering how to behave in a situation, especially if you feel tension, consider the three questions in the Defiance Compass.

THE DEFIANCE COMPASS

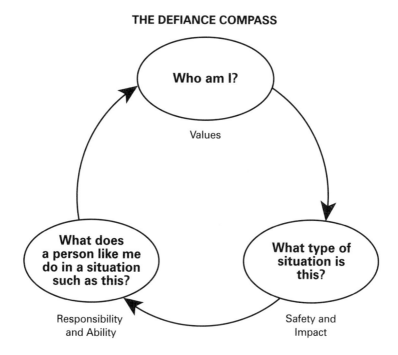

Self-connection means getting in touch with your core values.

It is *not:*

- Allegiance to a cause

- Membership in a party

It *is:*

- Understanding your own biases and rationalizations

- Seeing through your self-deceptions

- Knowing your true values

Take the time to reflect, articulate, and write down your values.

When you are about to take an action you are unsure about, ask yourself the following questions:

- *Is this me?*

- *Does this action represent my core values?*

Chapter 11: The Right Place and Time

For every moment of defiance, we can assess the environment by asking: *What type of situation is this?*

To answer it, ask yourself two further questions:

- *Is it safe for me to defy?* (safety)

- *Will this action make any difference?* (positive impact)

We cannot always control our environment, but we are often more effective than we think.

Small-scale defiance develops our skills, increases our confidence, and prepares us to recognize the right time and place to defy.

The **defiance empathy gap** represents the discrepancy between people's beliefs regarding the safety, ease, and effectiveness of defiance in a particular environment.

Asking for clarification can **raise the volume** on a situation—and when you raise the volume, you can begin to transform it. Asking "What do you mean by that?" or "Do I understand you correctly?" are non-threatening ways to make your discomfort known and gauge the safety of further defiance.

Chapter 12: The Superpower of Responsibility

We make decisions about defiance based on our answer to the question: *What does a person like me do in a situation like this?*

Answering this question involves assessing two main factors:

- Responsibility—*Is it my duty to defy? Do I have a responsibility to act on behalf of others or myself?*

- Ability—*How well-equipped am I for defiance? Do I have the confidence and skill to defy?*

Responsibility can enable us to tap into our defiant power. When we are responsible for others—our friends, our family—we often find it easier to defy.

Chapter 13: Level Up

Increase your *ability* to defy with **preparation and practice**:

- **Anticipate** situations that might call for defiance and reflect on past experiences.

- **Visualize** how you might act.

- **Practice** by scripting and role-playing. Saying defiant words out loud allows our mouths to get used to speaking them and our ears to get used to hearing them. Engage in "small-scale" defiance—small defiant acts can lead to more change than you might anticipate.

- **Repeat** as necessary.

Anticipating, visualizing, and *practicing* reduces surprise, which can disable defiance.

Hone your skills by considering past successful and not so successful defiance attempts and ask:

- What enabled your defiance?

- What disabled it?

- What were the barriers to your defiance, and how could you mitigate them?

Preparation and practice are critical to changing how we might behave in a situation. Imagine if the McDonald's managers had been forewarned about crank calls from a "police officer" asking them to strip search their employees.

Just as The Talk provides young people of color with a context for understanding specific stressful encounters with the police, other types of talks could help prepare people for a range of different situations. For example, senior academics could forewarn junior colleagues about situations like the one I encountered as a postdoc. As one senior woman academic said at a conference workshop called *Yet, She Persisted,* "There is nothing wrong with you, but there is plenty wrong with the system." We can remember that institutional failure is not individual failure.

Just as Kevin the police officer anticipated that he would face ethical challenges in his line of work, we too can prepare ourselves and each other to understand that certain encounters are likely to come up, and they are not our fault. Parents could help their teenage children understand specific pressures that they might face and what a True Yes and a True No actually are. We can learn the difference between consent, conscious compliance, and defiance. And we can do all this without self-judgment and shame.

Chapter 14: You Don't Have to Be Brave

There are many paths to defiance, but they all begin in the same place: one person connecting with their core values and then taking action; that action can be small.

Moral mavericks act on their core values, realize their responsibility to defy, and give their True Yes or their True No to try to create a better world.

Defiance can change the world. But it also changes us. It makes us *more* ourselves. When we lean into what makes us unique, we can become defiant in our own ways, connecting with our aspirational selves: with both who we are and who we want to be.

Conclusion: Dare to Defy

Give yourself the power of the pause.

Take a moment to step out of a challenging situation and reflect; walk away, talk to yourself, even take a deep breath before responding.

Whenever you might need to defy:

Pause.

- Remember your values: "Who am I?"

- Consider the situation and assess for safety and positive impact.

- Connect with your responsibility and ask, "What does a person like me do in a situation like this?"

DISCUSSION QUESTIONS

1. When you were growing up, what were you taught about compliance and defiance?

2. When did you first witness someone being defiant? Who was it, and what value were they expressing? Did you know the impact of their act of defiance? How did it affect you?

3. Can you remember a specific moment in your childhood when you first defied an expectation, order, or suggestion? What was it, and why did you defy? What did you learn from this experience?

4. Do you think of yourself as a defiant person? Why or why not?

5. A True Yes is an action or agreement that is in alignment with your values and satisfies the elements of consent—capacity, knowledge, understanding, freedom, and your explicit authorization. When was the last time you gave your True Yes? How did it feel, and what were the consequences?

6. Defiance is a way of saying "no" to something out of alignment with your true values. What is one guiding value in your life? Can you identify what or who in your life instilled that value in you—was it a particular incident or person, as it was for Trevor? What choices do you make based on this value?

7. What is your version of a "crocodile smile"—an outward manifestation of conscious compliance when outright defiance isn't the best option for you? When was the last time you used it?

8. What physical sensations do you experience when you're torn between compliance and defiance? Is there a specific tension or feeling that you notice within yourself?

9. Can you think of a time you took bad advice, even though you had reason to distrust the advice or the person giving it? What was the reason you followed that advice?

10. What is a tool or script you can use to "pause" in a high-pressure situation so that you can check in with your own values before making a decision?

11. Who or what do you feel the most responsible for?

12. How do you engage in your form of quiet defiance?

13. How do you "raise the volume" on something that goes against your values? Have you seen others do this?

14. What was the last small defiant win in your life?

15. Where and when do you demonstrate your defiance more easily and with less angst? What values do you go out of your way to safeguard? The answer to both these questions helps identify your "defiance superpower."

16. What are the challenges to your values that you are most likely to face? With these challenges in mind, how can you anticipate, prepare, and practice for defiance in your life?

ACKNOWLEDGMENTS

Bringing *Defy* to life has been an incredible experience, filled with challenges but blessed with rewards I could not have imagined. Transforming my ideas into a tangible book was a journey that could only be achieved with the support, guidance, and love of countless individuals.

At the heart of this voyage, as with every adventure I undertake, stands my family. My husband, Mark, and our son, Samuel, have been my unwavering pillars of support. Samuel, your limitless joy and the spark you bring to my life brighten every day. Mark, not only did you endure numerous drafts and discussions with patience and insight, but your steadfast belief in the importance of this work has been a constant source of encouragement. The many cups of Earl Grey tea you lovingly prepared have sustained me throughout the twilight hours and greeted me at dawn, each sip a gentle reminder of the quiet strength and love that has made this book possible.

To my parents, your love and guidance have profoundly shaped me. The values you instilled in me find expression on these pages. Your story forever lives within mine, a narrative of dependable love and resilience. And to my brothers and sister, your influence is woven throughout this book, a testament to our shared journey.

To the late Nagesh Gavirneni, my colleague turned cherished friend, your insightful feedback and support stay with me. Your absence is deeply felt, yet your wisdom continues to guide me. And to Onoso Imoagene, your astute comments helped shape this book. I am profoundly grateful to both of you for your willingness to delve into the entirety of the manuscript, sharing your thoughts and time with such openness.

Ethan Kross and Luc Wathieu, your guidance and support, enriched by our long discussions, have been instrumental in navigating the intricate process of book creation. Kerry Ann Rockquemore, your mentorship deepened my understanding of defiance, and your advice and encouragement steered me through the myriad complexities of this endeavor.

My agent, Rachel Neumann, enthusiastically embraced *Defy* from the outset. Rachel, your dedication and keen editing not only enhanced the ideas within this book but also elevated it to heights I had only hoped to reach. Your ongoing support is deeply valued.

I could not have found a better editor for *Defy* than One World's Elizabeth Méndez Berry. From our very first conversation, when you shared your personal connection to the book's theme, I knew I had found a kindred spirit. Your passionate advocacy and meticulous critiques have challenged and inspired me, pushing me to delve deeper into the essence of defiance and significantly shaping the narrative of the book.

The collective efforts of the entire team at One World and Penguin Random House have created fertile ground for *Defy* to thrive. Chris Jackson, Oma Beharry, London King, Avideh Bashirrad, Lulú Martínez, Carla Bruce, Stephanie Bowen, Zehra Kayi, Rachael Perriello Henry, and the many others behind the scenes who have contributed their expertise. The environment they've cultivated is one where creativity and passion can flourish, and for that, I am deeply grateful.

Special thanks are due to Jordan Jacks for his creative insights, assiduous research, and staunch optimism, as well as his infinite patience in grasping precisely what I aimed to convey through my writing. Likewise, Alyssa Knickerbocker's uncanny ability to pinpoint the perfect places for a chapter break has been nothing short of essential. Together, your support and companionship have made this process a much less lonely one.

I extend my appreciation to Caspian Dennis and Sandy Violette and my wonderful U.K. Bonnier team, as well as Camilla Ferrier and others at the Marsh Agency. Thank you for finding *Defy* a home across the globe, ensuring its message reaches far and wide.

I am especially indebted to every single person I interviewed for sharing their stories of defiance and conscious compliance, and to the researchers whose studies have informed my thoughts. Your contributions are the foundation upon which *Defy* stands.

To all the incredible individuals who lent their eyes and minds to refine my introduction: Melissa Thomas-Hunt, Brittany K. Barnett, Jamie Lyn Perry, Neil Anthony Lewis, Jr., Erika V. Hall, Lindsey Cameron, Shanique Brown, Ayana O, Bertrand J. Odom-Reed, Terri Francis, Laura Stone, Inés Hermida, and others, your thoughtful insights helped craft a narrative that is sensitive in both content and tone.

Heartfelt thanks to Adam Grant, Robert Cialdini, Jonah Berger, Barry Schwartz, Tina Payne Bryson, Katy Milkman, Adam Alter, Maria Konnikova, Annie Duke, Wendy Wood, Deborah Gruenfeld, Marc Schulz, Dolly Chugh, Sarah Kaplan, Sahara Byrne, Michèle Lamont, Edith Eger, Dames Sandra Dawson, and Diane Coyle, and the many others who graciously reviewed my book proposal and endorsed me as a debut author. Your belief in the potential of this work has been a source of great motivation.

To my colleagues, students, and collaborators, your engagement and contributions have been essential to this journey. And to *all* my friends—those with whom I've shared years of close companionship and those I've had the good fortune to meet along this path, including members of the Top Three Book Workshop and the Women in Organizational Behavior writing groups—your support and kindness have been extraordinary gifts. Each one of you, whether you realize it or not, has played a pivotal role in this process, and for that I am deeply grateful.

This acknowledgment merely scratches the surface of my gratitude. To everyone who has journeyed with me, offered wisdom, and inspired defiance, I sincerely thank you. Your impact extends far beyond the pages of this book, contributing to a world that values the freedom to defy.

NOTES

In this section, I have compiled a detailed list of notes and sources used to put this book together. While I hope this list will meet the needs of most readers, scientific knowledge evolves with time, which means some references may require future updates. If you notice any errors, please reach out to me via sunitasah.com/contact so I can address the issue. An updated list of endnotes will be maintained at *sunitasah.com/defy/endnotes*.

Introduction: Moving the Knee

xvii **"When you think of . . .":** Sir Charles Percy Snow, "Either-Or," *Progressive* 25, no. 2 (1961): 24–25.

xvii **On May 25, 2020:** The account of Alex Kueng's and Thomas Lane's actions on May 25, 2020, derives from body cam footage and from Robert Samuels and Toluse Olorunnipa, *His Name is George Floyd: One Man's Life and the Struggle for Racial Justice* (New York: Viking, 2022). The body cam footage is detailed in Haley Willis et al., "New Footage Shows Delayed Medical Response to George Floyd," *The New York Times,* August 11, 2020, sec. U.S., nytimes.com/2020/08/11/us /george-floyd-body-cam-full-video.html. I also used voice transcripts from Alex Kueng's body cam that were submitted to the Minnesota court as evidence, mncourts.gov/mncourtsgov/media/High-Profile-Cases/27-CR-20-12951-TKL /Exhibit407072020.pdf. Also see Harmeet Kaur and Nicole Chavez, "What We Know about the 3 Ex-Police Officers on Trial This Week for George Floyd's Death," *CNN,* January 24, 2022, cnn.com/2022/01/24/us/minneapolis-officers -background-george-floyd/index.html.

xviii **his grandfather had been a homicide detective:** Alex Woodward, "Thomas Lane: What to Know About Former Minneapolis Officer Facing Federal Charges in George Floyd's Killing," *The Independent,* January 24, 2022, sec. News, independent.co.uk/news/world/americas/crime/thomas-lane-trial-george-floyd -b1999162.html.

xviii **Kueng had joined the police department:** This information on J. Alexander Kueng's upbringing and motivations derives from Kim Barker, "The Black Officer Who Detained George Floyd Had Pledged to Fix the Police," *The New York Times,* June 27, 2020, sec. U.S., nytimes.com/2020/06/27/us/minneapolis-police-officer -kueng.html.

xx   ***ionizing radiation:*** Fred A. Mettler et al., "Effective Doses in Radiology and
     Diagnostic Nuclear Medicine: A Catalog," *Radiology* 248, no. 1 (July 2008):
     254–63. doi.org/10.1148/radiol.2481071451. Also see "Radiation Risk from
     Medical Imaging," *Harvard Health Publishing*, Harvard Medical School, Septem-
     ber 22, 2010, health.harvard.edu/cancer/radiation-risk-from-medical-imaging.

xx   ***the bedrock principles of medical ethics:*** There are four principles of medical
     ethics: beneficence (doing good), nonmaleficence (doing no harm), autonomy (the
     freedom to choose when able), and justice (ensuring fairness). For more, see Basil
     Varkey, "Principles of Clinical Ethics and Their Applications to Practice," *Medical
     Principles and Practice* 30, no. 1 (2021): 17–28.

xxi ***"it's the way things are done":*** Sunita Sah, "Investigations Before Examinations:
     'This Is How We Practice Medicine Here,'" *JAMA Internal Medicine* 175, no. 3
     (March 2015): 342–43, doi.org/10.1001/jamainternmed.2014.7549.

xxv ***hear the word "defy":*** The *Oxford English Dictionary* (*OED*) defines "defy" as "To
     challenge the power of; to set at defiance; to resist boldly or openly; to set at
     nought." *OED Online,* s.v. "Defy, v.1, 4.a." accessed December 11, 2023, oed.com
     /dictionary/defy_v1?tab=meaning_and_use#7217215.

xxv **disobedience *as:*** The *OED* defines "disobedience" as "The fact or condition of
     being disobedient; the withholding of obedience; neglect or refusal to obey;
     violation of a command by omitting to conform to it, or a prohibition by acting in
     defiance of it; an instance of this." *OED Online,* s.v. "Disobedience, n., a." accessed
     December 11, 2023, oed.com/dictionary/disobedience_n?tab=meaning_and_use
     #6599845.

Chapter 1: Wired to Comply

5   ***our brain's level of dopamine:*** Ethan S. Bromberg-Martin, Masayuki Matsumoto,
     and Okihide Hikosaka, "Dopamine in Motivational Control: Rewarding, Aversive,
     and Alerting," *Neuron* 68, no. 5 (December, 2010): 815–34, doi.org/10.1016/j
     .neuron.2010.11.022.

6   ***"she who has good conduct or behavior":*** *Dictionary of Sanskrit Names,* compiled
     by the Integral Yoga Institute (Buckingham, Virginia: Integral Yoga Publications,
     1989).

14   ***Milgram conducted his now infamous studies:*** Stanley Milgram, "Behavioral
     Study of Obedience," *The Journal of Abnormal and Social Psychology* 67, no. 4
     (October 1963): 371–78, doi.org/10.1037/h0040525; Stanley Milgram, *Obedience
     to Authority* (New York: Harper & Row, 1974). While most of the early experi-
     ments Milgram conducted were on men aged 20 to 50 years, of varying races and
     all occupational levels from unskilled workers to professionals, a subsequent experi-
     ment of his focused entirely on women participants, and found strikingly similar
     results. [See "Experiment 8," summarized in *Obedience to Authority*, 62–63.]

16   ***"In a naïve moment some time ago . . .":*** Letter by Stanley Milgram, quoted in
     Thomas Blass, *The Man Who Shocked the World: The Life and Legacy of Stanley
     Milgram* (New York: Basic Books, 2004), 100.

17   **These "obedient" people:** When I read Milgram's studies, I saw that the experi-

ments were not only about our obedience to authority. His book also captured information on people's differing reactions to unethical orders and details on those participants who managed to defy the experimenter. Also see Matthew M. Hollander, "The Repertoire of Resistance: Non-Compliance with Directives in Milgram's 'Obedience' Experiments," *British Journal of Social Psychology* 54, no. 3 (2015): 425–44, doi.org/10.1111/bjso.12099, for an analysis of different types of resistance displayed by participants in Milgram's experiments.

Chapter 2: Tension Is Your Strength

21 *This scenario is modeled:* This teaching scenario was inspired by Jack Brittain and Sim Sitkin's popular case study, "Carter Racing" (originally 1987, updated 2006), currently available through Delta Leadership Incorporated, deltaleadership.com /store/case-studies-exercises/carter-racing-2015_1/.

22 *five Morton Thiokol engineers had concerns:* William Rogers et al., "Report of the Presidential Commission on the Space Shuttle *Challenger* Accident" (Washington, D.C., June 6, 1986), 86–92, sma.nasa.gov/SignificantIncidents/assets/rogers _commission_report.pdf.

23 *"How the hell can you ignore this?":* Douglas Martin, "Roger Boisjoly, 73, Dies; Warned of Shuttle Danger," *The New York Times,* February 4, 2012, sec. U.S., nytimes.com/2012/02/04/us/roger-boisjoly-73-dies-warned-of-shuttle-danger .html.

23 *Bob Ebeling:* Howard Berkes, "*Challenger* Engineer Who Warned of Shuttle Disaster Dies," *NPR,* March 21, 2016, sec. America, npr.org/sections/thetwo-way /2016/03/21/470870426/challenger-engineer-who-warned-of-shuttle-disaster-dies.

23 *Allan McDonald:* Allan J. McDonald, *Truth, Lies, and O-Rings: Inside the Space Shuttle Challenger Disaster* (Gainesville: University Press of Florida, 2012).

25 *Expert intuition:* Daniel Kahneman and Gary Klein, "Conditions for Intuitive Expertise: A Failure to Disagree," *American Psychologist* 64, no. 6 (2009): 515–26, doi.org/10.1037/a0016755.

28 *Jeffrey Wigand:* Marie Brenner, "Jeffrey Wigand: The Man Who Knew Too Much," *Vanity Fair,* May 1996, vanityfair.com/magazine/1996/05/wigand199605; *Jeffrey Wigand: The Big Tobacco Whistleblower,* interview with Mike Wallace, *60 Minutes,* February 4, 1996, youtube.com/watch?v=1_-Vu8LrUDk.

29 *Wigand was sued:* Brenner, "Jeffrey Wigand."

31 *they reached a historic settlement:* "The Master Settlement Agreement," National Association of Attorneys General, accessed December 15, 2023, naag.org/our -work/naag-center-for-tobacco-and-public-health/the-master-settlement -agreement/; "Tobacco Master Settlement Agreement," *Wikipedia,* September 13, 2023, en.wikipedia.org/w/index.php?title=Tobacco_Master_Settlement _Agreement.

32 *"I don't think I've been this happy":* Jeffrey Wigand, interview with Mike Wallace, "60 Minutes Follows up with Wigand," *CBS News,* January 13, 2005. Video: cbsnews.com/video/60-minutes-follows-up-with-wigand/; Summary transcript: cbsnews.com/news/battling-big-tobacco/.

Chapter 3: Know Your True Yes

35 *April 9, 2004:* "Restaurant Shift Turns into Nightmare," *ABC News,* November 9, 2005, abcnews.go.com/Primetime/story?id=1297922; Andrew Wolfson, "A Hoax Most Cruel: How a Caller Duped McDonald's Managers into Strip-Searching a Worker," *Louisville Courier Journal,* October 9, 2005, courier-journal.com/story /news/investigations/2022/05/05/strip-search-hoax-kentucky-mcdonalds-fake -officer-scam/9598367002/; Maira Ansari, "McDonald's Calls Witnesses in Strip Search Trial," *Wave 3 News,* September 25, 2007, wave3.com/story/7127864 /mcdonalds-calls-witnesses-in-strip-search-trial.

38 *between 1994 and 2004:* McDonald's Corp. v. Ogborn, 309 S.W.3d 274 (Ky. Ct. App. 2010).

39 *Consent is radically different from compliance:* Vanessa Bohns et al., "The Farce of Consent: Psychological Factors that Challenge the Notion of Voluntary Consent," *Academy of Management Annual Meetings Proceedings* 2019, no. 1 (August 2019), doi.org/10.5465/AMBPP.2019.11214symposium.

39 *medical definition of "informed consent":* For overviews of the relevant medical and legal definitions of consent, see Thomas Grisso and Paul S. Appelbaum, *Assessing Competence to Consent to Treatment: A Guide for Physicians and Other Health Professionals* (Oxford University Press, 1998); Stephen Joffe and Robert D. Truog, "Consent to Medical Care: The Importance of Fiduciary Context," in *The Ethics of Consent: Theory and Practice,* ed. Franklin G. Miller and Alan Werthheimer (Oxford University Press, 2009); Sunita Sah, "Ethical Considerations for Obtaining Informed Consent: Insights from Psychology," in *Improving Consent and Response in Longitudinal Studies of Aging: Proceedings of a Workshop,* ed. Brian Harris-Kojetin (Washington, D.C.: National Academies Press, 2022), doi.org/10.17226/26481.

41 *What happened that day:* The McDonald's hoax case has a long legal aftermath. Both Summers and Nix were charged with crimes related to what happened that day. Summers entered an Alford plea for a misdemeanor charge of criminal imprisonment—essentially, she maintained her innocence while acknowledging that there was sufficient evidence to convict her. Given probation, she later sued McDonald's, claiming that the corporation had failed to train its managers sufficiently to respond to hoaxes. Nix pleaded guilty to sexual abuse, sexual misconduct, and unlawful imprisonment, and was sentenced to five years in prison.

Summers ended their engagement the night of the hoax, after she watched the security video of Nix with Ogborn.

Ogborn abandoned her plans to go to college, suffered from PTSD and recurrent panic attacks, and needed years of therapy. She eventually sued McDonald's for $200 million in damages and was awarded a significant settlement. But the psychological scars affect her to this day.

As for the hoaxer, "Officer Scott"?

Detectives in multiple states were able to trace the hoax calls in several incidences to calling cards bought at a single Wal-Mart in Florida. Through security camera footage, they identified a man they believed to be Officer Scott.

The suspect, a prison guard, had long dreamed of a career in law enforcement. Detectives found police paraphernalia in his trailer, along with job applications and study guides for the police academy entrance exam he was never able to pass. They believed his hoax calls were an attempt to act out fantasies of control, authority, and domination.

But because Officer Scott did not physically abuse any of the victims himself—because all of his manipulation was done on the phone—he was only charged with impersonating a police officer and soliciting sodomy. And even then, a Kentucky jury was unconvinced that the largely circumstantial evidence could definitively tie the suspect to the calls. The security camera footage was blurry. There was no recording of his voice on the line, instructing Summers and Nix. Owning the calling cards couldn't prove beyond a reasonable doubt that the same man had made the calls.

In the end, the man on the end of the phone line was never held responsible or punished.

See Amelia Beamer, "Who Is Louise Ogborn and Where Is She Now?," *The US Sun,* December 15, 2022, the-sun.com/news/6920053/who-is-louise-ogborn-and -where-is-she-now/; Associated Press, "Acquittal in Hoax Call That Led to Sex Assault," *NBC News,* October 31, 2006, nbcnews.com/id/wbna15504125; Brett Barrouquere, "Hoax Victim Testifies Against McDonald's," *The Oklahoman,* September 20, 2007, oklahoman.com/story/news/2007/09/20/hoax-victim -testifies-against-mcdonalds/61710898007/.

42 **Do you have a boyfriend?:** Julie A. Woodzicka and Marianne LaFrance, "Real Versus Imagined Gender Harassment," *Journal of Social Issues* 57, no. 1 (2001): 15–30.

42 ***People often experience "empathy gaps":*** Leaf van Boven et al., "Chapter Three—Changing Places: A Dual Judgment Model of Empathy Gaps in Emotional Perspective Taking," in *Advances in Experimental Social Psychology*, ed. James M. Olson and Mark P. Zanna, vol. 48 (Academic Pressure, 2013), 117–171, doi.org/10.1016/B978-0-12-407188-9.00003-X. Also see chapter 11 for more on empathy gaps.

42 ***high-power people:*** Marvin A. Hecht and Marianne LaFrance, "License or Obligation to Smile: The Effect of Power and Sex on Amount and Type of Smiling," *Personality and Social Psychology Bulletin* 24, no. 12 (December 1, 1998): 1332–42, doi.org/10.1177/01461672982412007.

44 ***a succession of women sued:*** Benjamin Wallace, "Bikram Feels the Heat," *Vanity Fair,* December 19, 2013, vanityfair.com/style/scandal/2014/01/bikram -choudhury-yoga-sexual-harassment.

44 ***"Good night, sir":*** *Bikram: Yogi, Guru, Predator,* directed by Eva Orner, (Pulse Films, 2019); Suparna Sharma, "TIFF 2019: Bikram Yogi and the Easy Asanas of Sexual Abusers," *Deccan Chronicle,* September 14, 2019, deccanchronicle.com /entertainment/bollywood/140919/tiff-2019-bikram-yogi-and-the-easy-asanas-of -sexual-abusers.html.

44 ***Armchair observers often attack victims:*** Boven et al., "Chapter Three— Changing Places."

45 *a 2014 illustrated essay:* "Trigger Warning: Breakfast," *The Nib,* July 8, 2014, thenib.com/trigger-warning-breakfast.

45 *the other reality was devastating:* This writer's experience is hardly unique—a 2015 meta-analysis of twenty-eight studies examining unwanted sexual experiences, obtained through force, threat of force, or incapacitation of the victim, in female survivors aged fourteen or older found that more than half of the survivors did not acknowledge what they experienced as "rape." Researchers use the term *unacknowledged rape* to describe this widespread phenomenon. But the data shows that this may be changing—in the aftermath of the #MeToo movement, which helped many victims understand the truth of what happened to them, a 2021 study of more than 2,500 college students found that even though the prevalence of sexual assault had likely remained unchanged, survivors were more likely to label their experiences of unwanted sexual experiences as "sexual assault." See Laura C. Wilson and Katherine E. Miller, "Meta-Analysis of the Prevalence of Unacknowledged Rape," *Trauma, Violence, & Abuse* 17, no. 2 (April 1, 2016): 149–159, doi.org/10.1177/1524838015576391; and Anna E. Jaffe, Ian Cero, and David Dilillo, "The #MeToo Movement and Perceptions of Sexual Assault: College Students' Recognition of Sexual Assault Experiences Over Time," *Psychology of Violence* 11, no. 2 (March 2021): 209–18, doi.org/10.1037/vio0000363.

45 *"non-promotable tasks":* Linda Babcock et al., *The No Club: Putting a Stop to Women's Dead-End Work* (New York: Simon and Schuster, 2022), 17–53.

47 *Thomas Simms:* Ansari, "McDonald's Calls Witnesses in Strip Search Trial."

Chapter 4: Break Free from Influence

54 *Although some Western cultures:* LuMing Robert Mao, "Beyond Politeness Theory: 'Face' Revisited and Renewed," *Journal of Pragmatics* 21, no. 5 (May 1, 1994): 451–86, doi.org/10.1016/0378-2166(94)90025-6.

54 *ideal* independent *view:* Hazel R. Markus and Shinobu Kitayama, "Culture and the Self: Implications for Cognition, Emotion, and Motivation," *Psychological Review* 98, no. 2 (1991): 224–53, doi.org/10.1037/0033-295X.98.2.224.

55 **insinuation anxiety:** Sunita Sah, "Why You Find It So Hard to Resist Taking Bad Advice," *The Los Angeles Times,* October 22, 2019, latimes.com/opinion/story /2019-10-29/advice-neuroscience-psychology-social-pressure-research; Sunita Sah, "Insinuation Anxiety and the Desire to Save the Other's Face: A Conceptualization and Model," working paper, 2024.

57 *According to a 2012 survey:* Nadine Bienefeld and Gudela Grote, "Silence That May Kill," *Aviation Psychology and Applied Human Factors* 2, no. 1 (January 2012): 1–10, doi.org/10.1027/2192-0923/a000021.

57 *A 2005 survey:* David Maxfield et al., *Silence Kills: The Seven Crucial Conversations for Healthcare* (Report commissioned by Vitalsmarts, Provo, UT, 2005). See summary at faculty.medicine.umich.edu/sites/default/files/resources/silent _treatment.pdf.

58 *In her book* **How Professors Think:** Michèle Lamont, "Pragmatic Fairness: Customary Rules of Deliberation," in *How Professors Think: Inside the Curious*

World of Academic Judgment (Cambridge, MA: Harvard University Press, 2009), 144–45.

59 ***In a series of experiments:*** Sunita Sah, George Loewenstein, and D. M. Cain, "Insinuation Anxiety: Concern That Advice Rejection Will Signal Distrust after Conflict of Interest Disclosures," *Personality and Social Psychology Bulletin* 45, no. 7 (2019): 1099–1112, doi.org/10.1177/0146167218805991.

61 ***expected more benevolence from their "advisor":*** "Benevolent sexism" is a set of attitudes that are seemingly positive (e.g., protective paternalism) yet reinforce women's subordinate status. See Peter Glick and Susan T. Fiske, "An Ambivalent Alliance: Hostile and Benevolent Sexism as Complementary Justifications for Gender Inequality," *American Psychologist* 56, no. 2 (2001), 109–18, doi.org/10.1037/0003-066X.56.2.109.

62 ***an experiment on board a mobile behavioral:*** Sunita Sah, George Loewenstein, and D. M. Cain, "The Burden of Disclosure: Increased Compliance with Distrusted Advice," *Journal of Personality and Social Psychology* 104, no. 2 (2013): 289–304, doi.org/10.1037/a0030527.

Chapter 5: Reclaim Your Power

66 ***In my medical scenario studies:*** Sunita Sah, George Loewenstein, and D. M. Cain, "Insinuation Anxiety: Concern That Advice Rejection Will Signal Distrust after Conflict of Interest Disclosures," *Personality and Social Psychology Bulletin* 45, no. 7 (2019): 1099–1112, doi.org/10.1177/0146167218805991.

66 ***And in the lottery studies:*** Sunita Sah, George Loewenstein, and D. M. Cain, "The Burden of Disclosure: Increased Compliance with Distrusted Advice," *Journal of Personality and Social Psychology* 104, no. 2 (2013): 289–304, doi.org/10.1037/a0030527.

67 ***Third-person self-talk:*** Ethan Kross, *Chatter: The Voice in Our Head, Why It Matters, and How to Harness It* (New York: Crown, 2021).

67 ***simply naming the feeling:*** Jared B. Torre and Matthew D. Lieberman, "Putting Feelings into Words: Affect Labeling as Implicit Emotion Regulation," *Emotion Review* 10, no. 2 (April 1, 2018): 116–24, doi.org/10.1177/1754073917742706.

Chapter 6: Find Your True No

77 ***On December 1, 1955:*** This chapter's account of Rosa Parks's story draws primarily from Jeanne Theoharis, *The Rebellious Life of Mrs. Rosa Parks* (Boston: Beacon Press, 2013). I also consulted the following texts: Henry Hampton and Steve Fayer, *Voices of Freedom: An Oral History of the Civil Rights Movement from the 1950s Through the 1980s* (New York: Random House, 1991); Bettye Collier-Thomas and V. P. Franklin, *Sisters in the Struggle: African American Women in the Civil Rights-Black Power Movement* (New York: NYU Press, 2001); Martha S. Jones, *Vanguard: How Black Women Broke Barriers, Won the Vote, and Insisted on Equality for All* (New York: Basic Books, 2020); and Rosa Parks and Jim Haskins, *Rosa Parks: My Story* (New York: Penguin, 1999).

77 *"Are you going to stand up?":* Theoharis, *The Rebellious Life of Mrs. Rosa Parks*, 63.

78 *"People always say":* Rosa Parks, *Rosa Parks: My Story* (New York: Dial, 1992), 132.

78 **A Los Angeles Times** *article:* " 'I'd Do It Again,' Says Rights Action Initiator," *Los Angeles Times,* December 16, 1965, documents.latimes.com/id-do-it-again-says -rosa-parks/.

78 *obituary in* **The New York Times:** Michael Janofsky, "Thousands Gather at the Capitol to Remember a Hero," *The New York Times,* October 31, 2005, sec. U.S., nytimes.com/2005/10/31/politics/thousands-gather-at-the-capitol-to-remember -a-hero.html.

78 *"a life history of being rebellious":* Theoharis, *The Rebellious Life of Mrs. Rosa Parks,* 1–16.

79 *her mother politely refused to move:* Theoharis, *The Rebellious Life of Mrs. Rosa Parks*, 51.

80 *pressuring the police . . . :* In the late 1960s, Rosa Parks wrote in an unpublished short story in the first person about a white man who had attempted to rape her when she was a teenager working for a wealthy family. She described how she cunningly persuaded the man not to assault her, by bringing up the very Jim Crow anti-miscegenation laws she would later help shatter.

"I was ready to die," she wrote, "but give my consent, never. Never. Never."

Although some suggest that this narrative might be a memoir, there's no mention of it in her autobiography or interviews, and it's believed by others that Parks penned it as an allegory, hinting at broader themes of dominance and defiance.

For more, see Theoharis, *The Rebellious Life of Mrs. Rosa Parks;* Ula Iinytzky, "Rosa Parks Essay Reveals Rape Attempt," *The Huffington Post,* July 29, 2011, sec. Culture, huffpost.com/entry/rosa-parks-essay-rape_n_912997; and Danielle L. McGuire, "The Maid and Mr. Charlie: Rosa Parks and the Struggle for Black Women's Bodily Integrity," in *U.S. Women's History: Untangling the Threads of Sisterhood,* ed. Leslie Brown, Jacqueline Castledine, and Anne Valk (New Bruns-wick, NJ: Rutgers University Press, 2017).

Parks's story is situated prior to her activism seeking justice for Black victims of sexual violence in the Jim Crow South—most notably Recy Taylor, a twenty-four-year-old Black woman in Alabama who was kidnapped and raped by six white men as she walked home from church. In 1944, Parks was sent by the NAACP to investigate Taylor's case, which ended with two grand juries refusing to indict the men, even though they confessed.

Parks's work on such important but underappreciated cases throughout the 1940s exemplifies her understanding of how interconnected racial and gender injustices were in the Jim Crow South. She knew, from an early age, that the personal was political—especially for Black women in a society that oppressed them for both their race and their gender.

For more on these subjects, see Theoharis, *The Rebellious Life of Mrs. Rosa Parks;* Danielle L. McGuire, *At the Dark End of the Street: Black Women, Rape, and Resistance—A New History of the Civil Rights Movement from Rosa Parks to the Rise of Black Power* (New York: Knopf Doubleday, 2011); "Hidden Pattern of Rape Helped Stir Civil Rights Movement," *NPR,* February 28, 2011, npr.org/templates /story/story.php?storyId=134131369.

80 ***"You better not hit me":*** Theoharis, *The Rebellious Life of Mrs. Rosa Parks,* 61.

81 ***"I thought of Emmett Till":*** Jesse Jackson, "Jesse Jackson Recalls Bus Boycott," interview with Ed Gordon, *News and Notes,* NPR, December 5, 2005, npr.org /templates/story/story.php?storyId=5039020.

81 ***in her poem "Rosa Parks":*** Nikki Giovanni, "Rosa Parks," collected in *Quilting the Black-Eyed Pea* (New York, HarperCollins, 2002). Accessed online via the Poetry Foundation: poetryfoundation.org/poems/90180/rosa-parks.

83 ***A progression through a series of stages:*** Milgram documented a similar sequence in participants who defied—of doubt, externalization, dissent, threat, and disobedience. See Stanley Milgram, *Obedience to Authority* (Harper & Row New York, 1974), 163.

84 ***participants who refused to administer shocks:*** Andre Modigliani and François Rochat, "The Role of Interaction Sequences and the Timing of Resistance in Shaping Obedience and Defiance to Authority," *Journal of Social Issues* 51, no 3 (1995): 107–23, doi.org/10.1111/j.1540-4560.1995.tb01337.x.

Chapter 7: The False Defiance Trap

87 ***On January 6, 2021, President Donald Trump:*** Alan Feuer et al., "Jan. 6: The Story So Far," *The New York Times,* June 9, 2022, sec. U.S., nytimes.com /interactive/2022/us/politics/jan-6-timeline.html.

88 ***Clayton Ray Mullins:*** Dan Barry, Alan Feuer, and Matthew Rosenburg, "90 Seconds of Rage," *The New York Times,* October 16, 2021, sec. U.S., nytimes.com /interactive/2021/10/16/us/capitol-riot.html.

88 ***pled guilty to the felony offense:*** District of Columbia United States Attorney's Office, "Kentucky Man Pleads Guilty to Assaulting Law Enforcement During Jan. 6 Capitol Breach," Department of Justice, September 6, 2023, justice.gov /usao-dc/pr/kentucky-man-pleads-guilty-assaulting-law-enforcement-during-jan -6-capitol-breach.

88 ***he was sentenced to thirty months:*** District of Columbia United States Attorney's Office, "Kentucky Man Sentenced to Prison for Assaulting Law Enforcement During Jan. 6 Capitol Breach," Department of Justice, January 30, 2024, justice .gov/usao-dc/pr/kentucky-man-sentenced-prison-assaulting-law-enforcement -during-jan-6-capitol-breach.

89 ***In one of the variations:*** Stanley Milgram, *Obedience to Authority* (Harper & Row New York, 1974), 116–21.

91 ***Algorithms on social media:*** Katherine J. Wu, "Radical Ideas Spread Through Social Media. Are the Algorithms to Blame?" *PBS,* March 28, 2019, pbs.org/wgbh /nova/article/radical-ideas-spread-social-media-algorithms/.

92 ***an expert on the Old Testament:*** Milgram, *Obedience to Authority,* 1974, 47–49.

92 ***the "proximity version":*** Milgram, 33–36.

94 ***"moral convictions":*** Linda J. Skitka, et al., "The Psychology of Moral Convic-

tion," *Annual Review of Psychology* 72, (2021): 347–66, doi.org/10.1146/annurev
-psych-063020-030612.

94 **On May 31, 2009 . . . :** Most of the information for this story comes from this
investigative piece: Judy L Thomas and David Klepper, "The Complex Life of
George Tiller," *Kansas City Star,* June 7, 2009, kansascity.com/news/state/kansas
/article230986418.html. I also consulted and drew from the following: Monica
Davey, "Witness Tells of Doctor's Last Seconds," *The New York Times,* July 28,
2009, sec. U.S., nytimes.com/2009/07/29/us/29tiller.html; Ed Pilkington, "I Shot
US Abortion Doctor to Protect Children, Scott Roeder Tells Court," *The
Guardian,* January 29, 2010, sec. World News, theguardian.com/world/2010/jan
/28/scott-roeder-abortion-doctor-killer.

95 **at church that morning:** *State v. Roeder,* 300 Kan. 901, 336 P.3d 831 (Kan. 2014).

Chapter 8: Who Gets to Defy?

Many of the stories in this chapter and the next are from interviews I conducted with my
business or executive students and their connections. To protect their privacy, I have given
these people pseudonyms and changed other identifying details.

101 **more likely to be killed:** Gabriel L. Schwartz & Jaquelyn L. Jahn, "Mapping Fatal
Police Violence across U.S. Metropolitan Areas: Overall Rates and Racial/Ethnic
Inequities, 2013–2017," *PLoS ONE* 15, no. 6 (June 24, 2020), doi.org/10.1371
/journal.pone.0229686.

103 **Beyoncé put it in "Flawless":** Kate Torgovnick May, "Beyoncé Samples Chimamanda Ngozi Adichie's TEDx Message on Suprise Album," *TEDBlog,* December 13, 2013, blog.ted.com/beyonce-samples-chimamanda-ngozi-adichies-tedx
-message-on-surprise-album/; Chimamanda Ngozi Adichie, "We Should All Be
Feminists," *TEDxEuston,* April 12, 2012, ted.com/talks/chimamanda_ngozi
_adichie_we_should_all_be_feminists.

104 **"Double Jeopardy?":** Joan C. Williams, "Double Jeopardy? An Empirical Study
with Implications for the Debates over Implicit Bias and Intersectionality,"
Harvard Journal of Law and Gender 37, (2014): 185–242.

104 **stereotypes of Asian women:** Virginia W. Wei, "Asian Women and Employment
Discrimination: Using Intersectionality Theory to Address Title VII Claims Based
on Combined Factors of Race, Gender and National Origin," *Boston College Law
Review* 37, no 4 (July 1, 1996): 771–812.

104 **able to defy loudly and publicly:** There is another category of defiant people who
are afforded little safety, security, or protection, and yet stand up anyway: people
like Sylvia Rivera, who are born into a world that does not accept them. Sylvia,
assigned male at birth, was raised by her grandmother in New York City. She
began experimenting with clothing and makeup at a young age and was beaten up
for doing so. Simply being herself was an act of daily defiance, one that regularly
exposed her to violence, hatred, and discrimination. Homeless from the age of
eleven, she understood the struggle of people like her, who had been cut off from

their families, isolated from their communities, and thrust into poverty. In 1970, at age nineteen, along with Marsha P. Johnson, an African American self-identified drag queen and activist, Sylvia cofounded Street Transvestite (later Transgender) Action Revolutionaries (STAR) to support homeless gay and transgender youth. In the ensuing years, she became an iconic figure in the fight against the exclusion of transgender people, especially transgender people of color, from the larger movement for gay rights. See Emma Rothberg, "Sylvia Rivera," National Women's History Museum, March 2021, womenshistory.org/education-resources /biographies/sylvia-rivera.

105 *"That's because of my parents":* Mindy Kaling, quoted at the 2015 Sundance Film Festival's *Power of Story: Serious Ladies* panel (Mindy Kaling, Lena Dunham, Jenji Kohan, and Kristen Wiig). See Jarett Wieselman, "The Secret Behind Mindy Kaling's Success Is Incredible," *BuzzFeed News,* January 24, 2015, buzzfeednews .com/article/jarettwieselman/mindy-kaling-shared-a-truly-amazing-secret-to-her -success.

105 *Steve Jobs:* John Simons, "Management Style: When Is It Viewed as Abusive?" *The Wall Street Journal,* August 9, 2016, wsj.com/articles/management-style-when-is-it -viewed-as-abusive-1470752281; Robert J. Bies, Thomas M. Tripp, and Debra L. Shapiro, "Abusive Leaders or Master Motivators? 'Abusive' Is in the Eye of the Beholder," in *Understanding the High Performance Workplace: The Line between Motivation and Abuse,* ed. Neal M. Ashkanasy, Rebecca J. Bennett, and Mark J. Martinko, SIOP Organizational Frontiers Series (New York: Routledge/Taylor & Francis Group, 2016), 252–76.

105 *Natalie Portman's remarks:* Erin Nyren, "Natalie Portman's Step-by-Step Guide on How to Topple the Patriarchy," *Variety,* October 12, 2018, variety.com/2018 /film/news/natalie-portman-gender-parity-power-of-women-1202978582/.

108 *she suffered for decades:* Jeanne Theoharis, *The Rebellious Life of Mrs. Rosa Parks* (Boston: Beacon Press, 2013), 116–164.

109 *"Yes, I'd do it again":* "'I'd Do It Again,' Says Rights Action Initiator," *Los Angeles Times,* December 16, 1965, documents.latimes.com/id-do-it-again-says-rosa -parks/.

110 *Nelson Mandela:* South African History Online, "Defiance Campaign 1952," accessed December 29, 2023, sahistory.org.za/article/defiance-campaign-1952.

110 *Mahatma Gandhi leading thousands:* Kenneth Pletcher, "Salt March | Indian History," *Encyclopedia Britannica,* November 23, 2023, britannica.com/event/Salt -March.

112 *There is a cost:* Diane S. Berry and James W. Pennebaker, "Nonverbal and Verbal Emotional Expression and Health," *Psychotherapy and Psychosomatics* 59, no. 1 (1993): 11–19, doi.org/10.1159/000288640.

Chapter 9: Quiet Defiance

120 *In one version of Milgram's:* Stanley Milgram, *Obedience to Authority* (New York: Harper & Row, 1974), 59–62.

124 **In a 2000 survey:** Neal Trautman, "Police Code of Silence Facts Revealed," research conducted by The National Institute of Ethics and presented at the Annual Conference for the International Association of Chiefs of Police, Legal Officers Section, 2000, aele.org/loscode2000.html.

125 **Walking with the Devil:** Michael W. Quinn, *Walking with the Devil: The Police Code of Silence: What Bad Cops Don't Want You to Know and Good Cops Won't Tell You* (Chicago: Quinn and Associates, 2005), 39–41.

127 **medical activism:** Matthew K. Wynia et al., "Medical Professionalism in Society," *New England Journal of Medicine* 341, no. 21, (November 18, 1999): 1612–16, doi.org/10.1056/NEJM199911183412112.

Chapter 10: Who Am I?

136 **doctors are not neutral observers:** Matthew K. Wynia et al., "Medical Professionalism in Society," *New England Journal of Medicine* 341, no. 21, (November 18, 1999): 1612–16, doi.org/10.1056/NEJM199911183412112; Russell L. Gruen, Steven D. Pearson, and Troyen A. Brennan, "Physician-Citizens—Public Roles and Professional Obligations," *JAMA* 291, no. 1, (January 7, 2004): 94–98, doi.org/10.1001/jama.291.1.94.

136 **higher incomes and education levels:** Ezekiel J. Emanuel, "Enhancing Professionalism Through Management," *JAMA* 313, no. 18 (May 12, 2015): 1799–1800, doi.org/10.1001/jama.2015.4336.

138 **stories of forced surgeries:** Caitlin Dickerson, Seth Freed Wessler, and Miriam Jordan, "Immigrants Say They Were Pressured into Unneeded Surgeries," *The New York Times,* September 29, 2020, sec. U.S., nytimes.com/2020/09/29/us/ice-hysterectomies-surgeries-georgia.html.

138 **As other physicians have proposed:** Emanuel, "Enhancing Professionalism Through Management."

139 **three questions:** James G. March, *Primer on Decision Making: How Decisions Happen,* (New York: Free Press, 1994).

141 **people who move abroad:** Hajo Adam et al., "The Shortest Path to Oneself Leads Around the World: Living Abroad Increases Self-Concept Clarity," *Organizational Behavior and Human Decision Processes* 145, (March 1, 2018): 16–29, doi.org/10.1016/j.obhdp.2018.01.002.

146 **The act of writing out:** Tracy Epton et al., "The Impact of Self-Affirmation on Health-Behavior Change: A Meta-Analysis," *Health Psychology: Official Journal of the Division of Health Psychology, American Psychological Association* 34, no. 3 (March 2015): 187–96, doi.org/10.1037/hea0000116.

146 **There is also evidence:** Jennifer A. Gregg et al., "Impact of Values Clarification on Cortisol Reactivity to an Acute Stressor," *Journal of Contextual Behavioral Science* 3, no. 4 (October 1, 2014): 299–304, doi.org/10.1016/j.jcbs.2014.08.002.

147 **When Jeffrey Wigand:** Marie Brenner, "Jeffrey Wigand: The Man Who Knew Too Much," *Vanity Fair*, May, 1996, vanityfair.com/magazine/1996/05/wigand199605.

Chapter 11: The Right Place and Time

151 *it takes about:* Michael Shapiro, "Super Bowl National Anthem: Average Duration Time of Pregame Song," *Sports Illustrated,* February 3, 2019, si.com/nfl /2019/02/03/super-bowl-liii-national-anthem-average-length-gladys-knight.

151 *a local beat reporter:* Jennifer Lee Chan, whose photo—the first to show Kaepernick kneeling—was originally posted to Twitter on August 26, 2016: @jenniferleechan, "Colin Kaepernick did not stand during the National Anthem— my picture provides proof," *Twitter*, August 27, 2016, twitter.com/jenniferleechan /status/769446898176495616. Reposted four years later: @jenniferleechan, "Four years ago today I took this photo, not knowing the full scale of what the importance and significance of it would be. It is the only photo of Colin Kaepernick sitting during the National Anthem. After speaking to Nate Boyer prior to the next game in San Diego, he knelt," *Twitter*, August 26, 2020, twitter.com /jenniferleechan/status/1298735998390550529.

151 *"I am not going to stand up . . .":* Colin Kaepernick, as told to Steve Wyche: "Colin Kaepernick Explains Why He Sat During National Anthem," NFL.com, August 27, 2016, nfl.com/news/colin-kaepernick-explains-why-he-sat-during -national-anthem-0ap3000000691077.

152 *"When there's significant change . . .":* Chris Biderman, "Transcript: Colin Kaepernick Addresses Sitting during National Anthem," *Niners Wire,* August 28, 2016, ninerswire.usatoday.com/2016/08/28/transcript-colin-kaepernick-addresses -sitting-during-national-anthem/.

152 *meant no disrespect:* Associated Press, "Colin Kaepernick Joined by Eric Reid in Kneeling for National Anthem Protest," *The Guardian*, September 1, 2016, sec. Sport, theguardian.com/sport/2016/sep/01/colin-kaepernick-eric-reid-kneel -national-anthem-protest-san-francisco-49ers.

153 *he left the 49ers:* Nunzio Ingrassia, "GM John Lynch Says 49ers Would Have Cut Colin Kaepernick If He Hadn't Opted Out," *FOX Sports,* June 1, 2017, foxsports .com/stories/nfl/gm-john-lynch-says-49ers-would-have-cut-colin-kaepernick-if-he -hadnt-opted-out.

155 *I interviewed fifty nurses:* Hyunsun Park, Sunita Sah, and Subrahmaniam Tangirala, "The Voice Empathy Gap: How Employees and Managers Hold Differing Beliefs About Lack of Voice and Interventions to Close the Gap," working paper, 2024.

156 *lack of psychological safety:* Amy C. Edmondson and Zhike Lei, "Psychological Safety: The History, Renaissance, and Future of an Interpersonal Construct," *Annual Review of Organizational Psychology and Organizational Behavior* 1, no. 1 (2014): 23–43, doi.org/10.1146/annurev-orgpsych-031413-091305.

158 *I examined the different perceptions:* Catherine T. Shea and Sunita Sah, "Just Don't Ask: Different Perceptions Between Recruiters and Candidates for Asking Illegal Demographic Questions in Job Interviews," working paper, 2024.

158 *educating managers:* Park, Sah, and Tangirala, "The Voice Empathy Gap: How

Employees and Managers Hold Differing Beliefs About Lack of Voice and
Interventions to Close the Gap."

159 **Rachael Denhollander liked Dr. Nassar:** Rachael Denhollander, *What Is a Girl
Worth?: My Story of Breaking the Silence and Exposing the Truth about Larry Nassar
and USA Gymnastics* (Carol Stream, Illinois: Tyndale House, 2019). See also Tim
Evans, Mark Aleisa, and Marisa Kwiatkowski, "USA Gymnastics Doctor Accused
of Abuse," *The Indianapolis Star,* September 12, 2016, indystar.com/story/news
/2016/09/12/former-usa-gymnastics-doctor-accused-abuse/89995734/; Juliet
Macur, "In Larry Nassar's Case, a Single Voice Eventually Raised an Army," *The
New York Times,* January 25, 2018, sec. Sports, nytimes.com/2018/01/24/sports
/rachael-denhollander-nassar-gymnastics.html; *Athlete A* (Actual Films, 2020),
actualfilms.net/films/athletea/.

160 **Sixteen years later:** Matt Mencarini, "She Exposed Her Secrets to Put Away
Sexual Predator Larry Nassar. But Her Work Isn't Done," *Louisville Courier
Journal,* September 4, 2019, courier-journal.com/in-depth/news/2019/09/04
/rachael-denhollander-sacrifice-continues-after-accusing-usa-gymnastics-larry
-nassar/1919109001/.

160 **A BLIND EYE TO SEX ABUSE:** Marisa Kwiatkowski, Mark Alesia, and Tim
Evans, "A Blind Eye to Sex Abuse: How USA Gymnastics Failed to Report
Cases," *The Indianapolis Star,* August 4, 2016, indystar.com/story/news
/investigations/2016/08/04/usa-gymnastics-sex-abuse-protected-coaches
/85829732/.

162 **"I came as prepared as possible":** Rachael Denhollander, "Rachael Denhollander:
The Price I Paid for Taking On Larry Nassar," *The New York Times,* January 26,
2018, sec. Opinion, nytimes.com/2018/01/26/opinion/sunday/larry-nassar
-rachael-denhollander.html.

163 **"The people make the place":** Benjamin Schneider, "The People Make the Place,"
Personnel Psychology 40, no. 3 (September 1987): 437–53, psycnet.apa.org/doi/10
.1111/j.1744-6570.1987.tb00609.x.

164 **small wins:** Karl E. Weick, "Small Wins: Redefining the Scale of Social Problems,"
American Psychologist 39, no. 1 (1984): 40–49, doi.org/10.1037/0003-066X.39
.1.40.

165 **Curiosity can be:** Todd B. Kashdan et al., "Curiosity Protects Against Interpersonal Aggression: Cross-Sectional, Daily Process, and Behavioral Evidence,"
Journal of Personality 81, no. 1 (2013): 87–102, doi.org/10.1111/j.1467-6494
.2012.00783.x.

166 **some restaurant owners attempt to:** Eran Dayan and Maya Bar-Hillel, "Nudge to
Nobesity II: Menu Positions Influence Food Orders," *Judgment and Decision
Making* 6, no. 4 (June 2011): 333–42, doi.org/10.1017/S1930297500001947;
Jungkeun Kim et al., "Position Effects of Menu Item Displays in Consumer
Choices: Comparisons of Horizontal Versus Vertical Displays," *Cornell Hospitality
Quarterly* 60, no. 2 (May 1, 2019): 116–24, doi.org/10.1177/193896551877
8234.

Chapter 12: The Superpower of Responsibility

168 ***The morning of May 24, 2022:*** This account of Angeli Rose Gomez's experience is derived primarily from *The Wall Street Journal*'s reporting on the aftermath of the Uvalde shooting and interviews with Gomez from CBS and ABC affiliate television news. See Douglas Belkin, Rob Copeland, and Elizabeth Findell, "Uvalde Shooter Fired Outside School for 12 Minutes Before Entering," *The Wall Street Journal,* May 26, 2022, sec. U.S., wsj.com/articles/uvalde-residents-voice -frustration-over-shooting-response-11653588161; "Mother Who Ran into Texas School During Shooting Discusses Moments Inside," *CBS News,* June 2, 2022, cbsnews.com/video/mother-who-ran-into-texas-school-during-shooting-discusses -moments-inside/; Rebecca Lopez, "Uvalde Mom Who Ran into School to Save Kids Says Police Are Harassing Her for Speaking Out," *WFAA News,* July 6, 2022, wfaa.com/article/news/special-reports/uvalde-school-shooting/uvalde-mom-says -police-are-harassing-her-for-speaking-out/287-25084f74-f3f4-49e9-b68b -b945c2f34df3; Chelsea Torres, "Mom Who Saved Her Kids from Uvalde School Shooting Says Police Are Targeting Her," *KATU* News, June 27, 2022, katu.com /news/nation-world/mom-who-saved-her-kids-from-uvalde-school-shooting-says -police-are-targeting-her-texas-robb-elementary-gunman-salvador-ramos-law -enforcement-response-angeli-rose-gomez-first-responders-who-is-pete-arredondo -town-square-protests-elementary-school-massacre.

176 ***"The most common adjustment of thought":*** Stanley Milgram, *Obedience to Authority* (Harper & Row New York, 1974), 7–8.

176 ***"it is a fundamental mode":*** Milgram, 8.

177 ***Bruno Batta:*** Milgram, 45–47.

178 ***"the agentic state":*** Milgram, 132–34.

178 ***Jan Rensaleer:*** Milgram, 50–52.

179 ***In several studies I conducted:*** Kaitlin Woolley and Sunita Sah, "Kicking Yourself: Going Against Your Inclinations Leads to Greater Feelings of Control and Culpability," working paper, 2024.

181 ***After the gymnasts told their stories:*** Marlene Lenthang, "How USA Gymnastics Has Changed since the Larry Nassar Scandal," *ABC News,* July 26, 2021, abcnews .go.com/Sports/usa-gymnastics-changed-larry-nassar-scandal/story?id=78839442.

182 ***a pair of researchers:*** James Habyarimana and William Jack, "Results of a Large-Scale Randomized Behavior Change Intervention on Road Safety in Kenya," *Proceedings of the National Academy of Sciences* 112, no. 34 (August 10, 2015), doi .org/10.1073/pnas.1422009112.

Chapter 13: Level Up

190 ***our mouths to get used:*** From the Heroic Public Speaking workshop, 2019, heroicpublicspeaking.com.

192 ***My refusal might just create:*** I have been inspired by Mary C. Gentile's work on speaking up in organizations and the importance of practice. See her book *Giving*

Voice to Values: How to Speak Your Mind When You Know What's Right (New Haven: Yale University Press, 2010).

193 **"Investigations Before Examinations . . .":** Sunita Sah, "Investigations Before Examinations: 'This Is How We Practice Medicine Here,'" *JAMA Internal Medicine* 175, no. 3 (March 1, 2015): 342–43, doi.org/10.1001/jamainternmed.2014.7549.

193 **her own supportive article:** Deborah Grady, "Testing Before Seeing the Patient," *JAMA Internal Medicine* 175, no. 3 (2015): 343, doi.org/10.1001/jamainternmed.2014.7607.

195 **"Under duress, we do not rise . . .":** This quote is often attributed to Bruce Lee, but its actual source is believed to be Archilochus, a seventh-century B.C. Greek poet.

Chapter 14: You Don't Have to Be Brave

197 **On August 20, 2018:** This account of Greta Thunberg's school strike draws heavily from the book she and her family published in Sweden in 2018, and in the United States in 2020. See Greta Thunberg, et al., *Our House Is on Fire: Scenes of a Family and a Planet in Crisis* (New York: Penguin, 2020), 223–278.

199 **"I was diagnosed with Asperger's":** Greta Thunberg, "School Strike for Climate—Save the World by Changing the Rules," *TEDxStockholm,* 2018, ted.com/talks/greta_thunberg_school_strike_for_climate_save_the_world_by_changing_the_rules/transcript.

202 **Ziauddin Yousafzai, an activist:** Alex Clark, "Malala's Father, Ziauddin Yousafzai: 'I Became a Person Who Hates All Injustice,'" *The Guardian,* November 11, 2018, sec. Books, theguardian.com/books/2018/nov/11/malala-father-ziauddin-yousafzai-i-became-a-person-who-hates-all-injustice.

202 **"You are a child":** Malala Yousafzai with Christina Lamb, *I Am Malala: The Girl Who Stood Up for Education and Was Shot by the Taliban* (New York: Little, Brown, 2013), 141.

204 **a group of elementary school children:** Thunberg et al., *Our House Is on Fire,* 257.

205 **"I'm gonna do something":** Greta Thunberg, appearance on *The Russell Howard Hour,* Series 6, Episode 8, November 4, 2022, youtube.com/watch?v=QqkhpxguKeY&t=1326s.

207 **"It feels like many today":** KK Ottesen, "Greta Thunberg on the State of the Climate Movement and the Roots of Her Power as an Activist," *Washington Post,* December 27, 2021, washingtonpost.com/magazine/2021/12/27/greta-thunberg-state-climate-movement-roots-her-power-an-activist/.

Conclusion: Dare to Defy

213 **You can see both of them:** For a relevant still image excerpted from the body cam footage, see Joe Hernandez, "Read This Powerful Statement from Darnella Frazier, Who Filmed George Floyd's Murder," *NPR,* May 26, 2021, npr.org/2021/05/26/1000475344/read-this-powerful-statement-from-darnella-frazier-who-filmed-george-floyds-murd; Haley Yamada, "Newly Released Video Show Onlookers'

Reaction to George Floyd's Death," *ABC News,* August 14, 2020, abcnews.go.com /US/newly-released-video-show-onlookers-reaction-george-floyds/story?id= 72378256. Excerpts from Tou Thao's body cam video and the bystanders' cellphone videos appear in *Latest Body Camera Video Shows Onlookers' Horror during George Floyd Arrest,* 2020, youtube.com/watch?v=GbSjqT4Zy1g.

214 *"It wasn't right":* Testimony of Darnella Frazier, recorded at the trial of Derek Chauvin, Hennepin County Courthouse, Minneapolis, MN, March 29, 2021, excerpted in Bill Chappell, " 'It Wasn't Right,' Young Woman Who Recorded Chauvin and Floyd on Video Tells Court," *NPR,* March 30, 2021, sec. America Reckons with Racial Injustice, npr.org/sections/trial-over-killing-of-george-floyd /2021/03/30/982729306/it-wasnt-right-young-woman-who-recorded-chauvin -and-floyd-on-video-tells-court.

215 *"It changed how I viewed life":* Darnella Frazier, "1 Year Anniversary," *Instagram,* May 25, 2021, instagram.com/p/CPT5_oIBlie/.

216 *"Between stimulus and response":* Viktor E. Frankl, *Man's Search for Meaning* (Simon and Schuster, 1985).

217 *"It's been said":* Greta Thunberg, et al., *Our House Is on Fire: Scenes of a Family and a Planet in Crisis* (New York: Penguin, 2020), 230.

INDEX

Page numbers in italics refer to figures or tables

ABOUT THE AUTHOR

DR. SUNITA SAH is an award-winning professor at Cornell University and an expert in organizational psychology. She leads groundbreaking research on influence, authority, compliance, and defiance. A trained physician, she practiced medicine in the United Kingdom and worked as a management consultant for the pharmaceutical industry. She currently teaches executives, leaders, and students in healthcare and business. Dr. Sah is a sought-after international speaker and consultant, advisor to government agencies, and former Commissioner of the National Commission on Forensic Science. Her multidisciplinary research and analyses have been widely published in leading academic journals and media entities including *The New York Times, Los Angeles Times, Harvard Business Review,* and *Scientific American.* She lives with her husband and son in New York.

SunitaSah.com

ABOUT THE TYPE

This book was set in Garamond, a typeface originally designed by the Parisian type cutter Claude Garamond (c. 1500–61). This version of Garamond was modeled on a 1592 specimen sheet from the Egenolff-Berner foundry, which was produced from types assumed to have been brought to Frankfurt by the punch cutter Jacques Sabon (c. 1520–80).

Claude Garamond's distinguished romans and italics first appeared in *Opera Ciceronis* in 1543–44. The Garamond types are clear, open, and elegant.